ALSO BY RAYMOND SMULLYAN

Theory of Formal Systems

First Order Logic

Gödel's Incompleteness Theorems

Recursion Theory for Metamathematics

Diagonalization and Self-Reference

Set Theory and the Continuum Problem

The Tao Is Silent

What Is the Name of This Book?

This Book Needs No Title

The Chess Mysteries of Sherlock Holmes

The Chess Mysteries of the Arabian Knights

The Lady or the Tiger?

Alice in Puzzleland

5000 B.C.

To Mock a Mockingbird

Forever Undecided

Satan, Cantor, and Infinity

THE RIDDLE OF SCHEHERAZADE

THE *RIDDLE* OF *SCHEHERAZADE*

AND

Other Amazing Puzzles, Ancient & Modern

BY

RAYMOND SMULLYAN

A Harvest Book
Harcourt Brace & Company
SAN DIEGO NEW YORK LONDON

The Riddle of Scheherazade is reprinted by arrangement with Alfred A. Knopf, Inc.

Library of Congress Cataloging-in-Publication Data
Smullyan, Raymond M.
The riddle of Scheherazade and other amazing puzzles, ancient & modern/Raymond Smullyan.
p. cm.—(A Harvest book)
Originally published: New York: A.A. Knopf, 1997.
ISBN 0-15-600606-5
1. Mathematical recreations. 2. Logic puzzles. I. Title. II. Title: Riddle of Scheherazade.
[QA95.S4497 1998]
793.7'4—DC21 98-16195

Text set in Adobe Garamond
Printed in the United States of America
First Harvest edition 1998

A C E F D B

Contents

Preface

Book One of this volume begins where Edgar Allan Poe left off in his remarkable story "The Thousand-and-Second Tale of Scheherazade." In this tale, Poe paints a very different picture of the ultimate fate of Scheherazade than that given in the *Arabian Nights*! But I have gone him one better and leave you here a puzzle-tale that I believe will intrigue and amuse you, and which ends in a way that naturally leads to a new field called Coercive Logic, which is the start of Book Two. This is followed by some new logic puzzles, logic tricks and games, Gödelian puzzles, and ends with some *very* baffling paradoxes!

Again my thanks to my editor, Ann Close, and the production editor, Melvin Rosenthal, for their expert help.

Elka Park, New York
September 1996

Book One

THE GRAND QUESTION OF SCHEHERAZADE

Chapter I

The Source

It will be remembered that in the usual version of the *Arabian Nights*, a certain monarch, having reason to believe in the infidelity of his queen, not only had her put to death but vowed by his beard and the Prophet to espouse each night the most beautiful maiden in his dominions, and the next morning to deliver her to the executioner. This unexampled inhumanity went on for some time and spread a panic of consternation throughout the city. Instead of the praises and blessings the people had once lavished upon their monarch, they now poured curses on his head. However, the grand vizier's elder daughter, Scheherazade, cleverly mastered the situation by marrying the king (against the urgent advice of her father) and arranging that her sister Dinerzade sleep with her in the nuptial chamber. Shortly before daybreak, she began telling her sister a wondrous tale, which the king overheard. When the hour of execution came, the monarch was so curious to hear how the tale ended that he granted her a twenty-four-hour stay of execution. The next night she finished the story, but began another one, which she was unable to finish in time (*sic!*), and so the king granted her another day's stay of execution. This went on for a thousand and one nights, at the end of which the king either forgot his vow or got himself absolved of it, and hence not only spared Scheherazade but ceased carrying out his ferocious edict.

All this, according to the *Arabian Nights*. Now, Edgar Allan Poe has informed us in his remarkable story "The Thousand-and-Second Tale of Scheherazade" that the ending as reported in the *Arabian Nights* is simply not correct! In his own words:

> Having had occasion, lately, in the course of some Oriental investigations, to consult the *Tellmenow Isitsöornot*, a work which—is scarcely known at all, even in Europe, and which has never been quoted to my knowledge, by any American—. I was not a little astonished to discover that the literary world has hitherto been strangely in error respecting the fate of the vizier's daughter, Scheherazade, as that the *denouement* there given, if not altogether inaccurate, as far as it goes, is at least to blame in not having gone much further. For full information on this interesting topic, I must refer the inquisitive reader to the "Isitsöornot" itself; but, in the meantime, I shall be pardoned for giving a summary of what I there discovered.

Poe then goes on to tell us what happened on the thousand-and-second night: "My dear sister," said Scheherazade, "now that this odious tax is so happily repealed, I feel that I have been guilty of withholding from you and the king—the full conclusion of the history of Sinbad the sailor—." She then went on to describe one miracle after another—or rather what *seemed* in those days to be miracles, but which in our time are simply common scientific truths—for example, things (light) that travel at the speed of 186,000 miles a second. The king got more and more irritatedly skeptical as Scheherazade went on, until he finally said: "Stop! I can't stand that and I won't. You have already given me a dreadful headache with your lies. The day, too, I perceive is beginning to break—Upon the whole, you might as well get up and be throttled." And so poor Scheherazade was executed.

All this according to the *Isitsöornot*. But unknown to its author, whoever that may be, as well as to Edgar Allan Poe, this fascinating book, like the *Arabian Nights*, is also in error concerning the fate of Scheherazade! It was my good fortune to be allowed to read another oriental book of such a secret nature that I had to swear I would never reveal its title. I can tell you, though, its subtitle, which is "A Critique of the *Isitsöornot*." My source is by far the most reliable of all, and goes on to explain that almost everything in the aforementioned volume is correct, but the last sentence is sadly in error. And now, I am happy to be able to report to you the real truth of the entire situation. This truth is more amazing than anything told in either the *Arabian Nights* or the *Isitsöornot*, and reveals Scheherazade as a female of such fantastic logical ingenuity that she might well be the envy of the greatest logicians of our time! What really happened will now be related:

True, the king did say (in Arabic), "Upon the whole, you might as well get up and be throttled." But Scheherazade replied, "Whatever pleases Your Majesty pleases me, but I am really sad for your sake, not mine."

"Why for *my* sake?" asked the king.

"Because of the puzzles I had planned to entertain you with," replied Scheherazade.

"Puzzles?" said the king. "I love puzzles! Will you tell me some tonight if I stay your execution?"

"I will tell you some every night as long as it pleases Your Majesty to let me live," replied Scheherazade.

And so, on the thousand-and-third night, the puzzles began, and went on for many nights, at the conclusion of which comes the most amazing part of the entire story! The puzzles themselves range from very simple and tricky to subtle and complex, culminating in the *grand question* of Scheherazade, which may well be the most clever logic puzzle of all time!

I will now set forth the events exactly as recorded in my secret source. Many of the puzzles (though not the best ones) have come down through history and are rather well known. But I will include them all—partly for the sake of those readers who may not be familiar with them, but mainly out of fidelity to my secret source, which is certainly a document of historical importance in its own right and should be treated with respect.

Chapter II

In Which Is Related How Scheherazade Entertained the King on the Thousand-and-Third Night

1 ☽ What Is It?

"Auspicious King," began Scheherazade on this memorable night, "let me first ask you a riddle: What is it that's greater than Allah; the dead eat it; and if the living eat it, they die?"

"Now, just a minute!" said the king. "That riddle has no answer! Nothing is greater than Allah, and it's blasphemous to say anything else!"

"I'm not being blasphemous," replied Scheherazade, "and you have just answered the riddle."

"What in the name of the Prophet are you talking about?" asked the king.

What is the answer to Scheherazade's riddle? (Solutions are given at the end of the book, pp. 177 ff.)

2 ☽ What Are the Chances?

"That was clever," said the king, after hearing the answer. "Tell me another."

"With pleasure," replied Scheherazade. "Tell me this: Are there at least two Arabian people who have exactly the same number of Arabian friends?"

"Now, how should I know?" asked the king.

"Let me make one point clear," said Scheherazade. "I am using the word *friends* in the mutual sense—for example, Ali is a friend of Ahmed only if Ahmed is also a friend of Ali. The relation of friendship is to be understood as symmetrical. Also, I am not regarding a person as a friend of himself."

"That doesn't help," said the king. "I still have no way of knowing—nor have you. For all I know, there *might* be two Arabian people with the same number of Arabian friends, and I daresay there probably are. But just what the chances are, I cannot say—nor can you."

The king was mistaken in his last assertion, for Scheherazade *did* know exactly what the chances were, and so will you if you solve this puzzle. What are the chances?

3 ☽ How Did They Manage?

"Neat!" said the king after hearing the solution. "Tell me another."

"Very well," she said. "Two camels were facing in opposite directions. One was facing due east and the other was facing due west. How can they manage to see each other without walking, turning around, or even moving their heads?"

"Hmm!" said the king, "I guess there must have been a reflection."

"No," said Scheherazade, "this was in the middle of the desert, and there was no reflecting material for miles around."

"Hmm," said the king.

How did they manage?

4 ☽ Abdul the Jeweler

"Cute!" said the king. "Tell me another."

"Very well, Auspicious King," said Scheherazade. "The news has reached me that one day a customer brought into the shop of Abdul the Jeweler six chains, each of which had five links. He wanted the six chains to be joined into one large circular chain and inquired as to the cost. 'Well,' replied the jeweler, 'every link I cut open and close costs one piece of silver.' The question, Your Majesty, is how many pieces of silver are required for the job?"

The king gave the wrong answer. What is the correct one?

5 ☽ The Second Tale of Abdul the Jeweler

"That was also clever!" said the king, after Scheherazade explained the right answer. "Tell me another!"

"Well," replied Scheherazade, "one night a thief stole into Abdul's shop—"

"He should be drawn and quartered!" interrupted the king.

"True, Your Majesty," replied Scheherazade, "but, to get on with my story, the thief joyfully came across a pile of diamonds. His first thought was to take them all, but then his conscience bothered him, and he decided to content himself with only half."

"Hmm!" said the king.

"And so he took half of the diamonds and started to leave the shop."

"Ho!" said the king.

"But then he thought: 'I'll take one more,' which he did."

"Hoo!" said the king.

"And so he left the shop, having stolen half the diamonds plus one."

"Then what happened?" asked the king.

"Strangely enough, a few minutes later, a second thief entered the shop and took half of the remaining diamonds plus one. Then a third thief entered the shop and took half the remaining diamonds and one more. Then a fourth thief entered and took half the remainder and one more. Then a fifth thief entered, but took no diamonds since they were all gone."

"So what is the problem?" asked the king.

"The problem," she said, "is how many diamonds were in the pile to start with?"

"Now, how should I know?"

"It's not difficult to figure out," she replied.

How many diamonds were in the pile?

6 ☽ Two Other Versions

"There are two other versions of the story," said Scheherazade. "According to a second version, each of the first four thieves took half of what he found plus two, instead of plus one, and again the fifth thief found none. According to this version, how many diamonds were originally in the pile?

"The third version is the same as the second, except that the fifth thief found one diamond. If this version is correct, then how many diamonds did the first thief find?"

7 ☽ Abdul and the Ten Thieves

"Another time," said Scheherazade, "ten thieves stole into Abdul's shop. Some of them were armed and some were unarmed. The armed ones were those of senior rank. Anyway, they stole a bag of fifty-six pearls. When it came to dividing them up, each senior robber took six pearls, and each junior robber got five. How many of the robbers were senior?"

8 ☽ How Many?

"Here is a happier one," said Scheherazade. "One day a man brought in fifty-nine jewels to sell to Abdul. Some were emeralds and some were rubies. The emeralds were carried in bags, nine to a bag, whereas the rubies were carried four to a bag. How many of the jewels were rubies?"

9 ☽ A Simple One?

"That was not bad for a simple one," said the king. "Give me another simple one."

"Very well," replied Scheherazade. "Which is more, six dozen dozen, or a half dozen dozen?"

"Now, really!" exclaimed the king, angrily. "I didn't want one *that* simple. The answer is so obvious! Do you take me for a dunce?"

What is the answer?

10 ☽ Sinbad and Hinbad

"Try this one," said Scheherazade. "There were two friends named Sinbad and Hinbad—"

"Are you referring to Sinbad the Sailor?" asked the king.

"Not necessarily," replied Scheherazade. "Anyway, they each owned the same number of horses. How many should Sinbad give Hinbad so that Hinbad has six more than Sinbad?"

"That's obvious!" said the king.

What is the answer?

11 ☽ Sinbad

"My next one *is* about Sinbad the Sailor," said Scheherazade. "On one of the boats on which Sinbad sailed, there was a ladder hung over the side that contained six rungs. The rungs were spaced one foot apart. At low tide, the water came up to the second rung from the bottom. Then the water rose two feet. Which rung did the water then hit?"

"Obviously, the fourth from the bottom," said the king. "Why do you give me such ridiculously simple ones?"

Do you agree with the king's answer?

12 ☽ How Many Ponies?

"Here's a little arithmetical one," said Scheherazade. "A certain sheik owned many ponies. Someone once asked him how many he had, and he replied: 'If you add one fourth of the number to one third of the number, you will then have ten more than one half the number.'

"How many ponies did he have?"

13 ☽ The Pony Who Got Lost

"Not bad," said the king. "Tell me another."

"Very well. One day one of the smaller ponies got lost in the desert for five days. He walked a certain distance the first day, and on each of the other days he walked one mile more than he had on the previous day. At the end of the five days, he was back home exhausted, since he had walked altogether fifty-five miles.

"How many miles did he walk on the last day?"

14 ☽ The Magic Tree

"Try this one," said Scheherazade. "A certain tree doubled its height every day—"

"Now, how do you expect me to believe that?" asked the king.

"It was a magic tree," replied Scheherazade.

"Oh, in that case, all right," said the king.

"Now," continued Scheherazade, "it took one hundred days

for the tree to reach its full height. How many days did it take to reach half its full height?"

"Obviously, fifty days," replied the king.

Was the king right?

15 ☽ Another Magic Tree

"Here is a more interesting one," said Scheherazade. "There was another magic tree that on the first day increased its height by half, on the second day by a third, on the third day by a quarter, and so on. How many days did it take to grow to one hundred times its original height?"

"Very nice," said the king. "But enough puzzles for tonight. Dawn is practically here, and I need some sleep. Will you tell me some more tomorrow night if I stay your execution?"

"With pleasure," replied Scheherazade. And thus we come to the next night.

Chapter III

Wherein Is Related How Scheherazade Further Entertained the King on the Thousand-and-Fourth Night

On this night Scheherazade began in a jocular mood.

"Before I tell you more puzzles," she said, "would Your Highness like to hear a funny story?"

"Yes?" said the king.

"Well, Karim the Woodchopper went to a lumber camp to apply for a job.

" 'Around here, we chop trees,' said the foreman.

" 'That's just what I do,' said Karim.

" 'Oh yes?' said the foreman. 'Let's see how good you are. Here's an ax. Let's see how long it takes you to take down that tree over there.' Karim gave the tree one mighty stroke, and the tree was down.

" 'Remarkable!' said the foreman. 'Well, let's see how long it takes you to topple that real big one over there.' In two mighty strokes the tree was felled.

" 'Fantastic!' said the foreman. 'Of course, you are hired, but how did you ever learn to chop like that?'

" 'Oh,' replied Karim. 'I got plenty of experience in the Sahara forest.' "

"Just a minute," interrupted the king. "Don't you mean the Sahara *desert*?"

"That's just what the foreman said," replied Scheherazade. "He said, 'You mean the Sahara *desert*, don't you?'

" 'Oh yeah,' replied the woodchopper, 'it is *now*.' "

"Not bad," laughed the king. "Now give me a puzzle."

16 ☽ Hassan's Horses

"Very well," said Scheherazade. "A certain sheik named Hassan had eight horses. Four of them were white, three were black, and one was brown. How many of Hassan's horses can each say that it is the same color as another one of Hassan's horses?"

17 ☽ How Much?

"Try this one," said Scheherazade. "How much is one million divided by one fourth, plus fifty?"

18 ☽ Hassan's Horses Again

"In the puzzle I gave you about Hassan's horses," said Scheherazade, "if I had given you the additional information that these horses could all talk, would seven then have been the correct answer?"

"Then it would have been correct," said the king.

"I believe not," said Scheherazade.

Who was right, and why?

19 ☽ Hassan's Mule

"Enough silly tricks," shouted the king. "Give me a straight one!"

"All right," said Scheherazade. "Hassan also owned a mule. Someone once asked him the age of his mule. He quizzically replied, 'In another four years, he will be three times as old as he was four years ago.'

"How old was the mule?"

20 ☽ What Color?

"To continue with Hassan," said Scheherazade, "he once met three boys and told them about his mule. 'What color is he?' asked one.

" 'Well,' replied Hassan, 'let's play a little guessing game. I will tell you that he is either brown, black, or gray. Why don't you try making some guesses, and when we have enough, I'll tell you some things about the guesses, and we'll then see if you can deduce his color.'

" 'I'll guess that he is not black,' said one.

" 'I'll guess that he is either brown or gray,' said another.

" 'I'll guess that he is brown,' said the third.

" 'Hold it!' said Hassan. 'We have enough guesses. It so happens that at least one of you guessed right and at least one of you guessed wrong.' "

What color is Hassan's mule?

21 ☽ Hassan's Camels

"Good," said the king. "Tell me another."

"Very well," said Scheherazade. "Hassan also owned eight camels. In one unfortunate month, all but five of them died. How many were left?"

"Obviously, three," said the king. "Any dunce could tell you that!"

I agree with the king's last statement. Do you? If not, see my reasons in the solution.

22 ☽ How Many Wives?

"No more tricks!" said the king sternly. "Tell me a genuine one."

"Very well," laughed Scheherazade. "Here is a genuine, albeit quite simple, one. A certain man has one less wife than his older brother. The older brother, in turn, has one less wife than their uncle. The younger brother has only half as many wives as the uncle. How many wives does each of the three men have?"

23 ☽ How Tall Is the Plant?

"That was indeed simple," said the king. "Try me on another."

"Very well," said Scheherazade. "If a certain plant were three feet taller, then it would be twice as tall as it would be if it were half a foot less.

"How tall is the plant?"

24 ☽ How Tall Are the Flowers?

"Another," said the king.

"Well, there are two flowers—one red and one blue. The red one is seven inches taller than the blue one. If the red one were four inches shorter, then it would be twice as tall as the blue one. How tall is each?"

25 ☽ How Far?

"Another," said the king.

"Well, one day a pet cat walked away from its home at the rate of three miles an hour. Suddenly, she remembered it was dinner time, and so she trotted back twice as fast. Altogether she was gone fifteen minutes. How far did she go?"

26 ☽ How Many Mice?

"Another," said the king.

"Very well. This same cat was very good at catching mice. On the first day, she caught one third of the mice. On the next day, she caught one third of the remaining mice. On the third day, she caught one third of the remaining mice. On the fourth day, she caught the remaining eight mice. How many mice were there?"

27 ☽ Ali and His Pets

"Another," said the king.

"All right," said Scheherazade. "A boy named Ali owned

some cats and dogs. He had more cats than dogs. One day, an evil magician flew over his house and—"

"Just a minute," interrupted the king (who was very practical). "I didn't know magicians could fly!"

"Most of them don't," she replied, "but this one did."

"But how *could* he?" asked the king.

"Because he was a *flying* magician," she replied.

"Oh, that explains it," he said. "Go on!"

"Well, this evil magician flew over the house and magically transformed one of the cats into a dog. Imagine Ali's surprise when he awoke the next morning and found that he now had the same number of dogs as cats! The next night, a good magician flew over the house and transformed the dog back into a cat. So when Ali awoke the next morning, things were back to normal. However, on the third night, another evil magician flew over the house, and this time transformed one of Ali's dogs into a cat! When Ali awoke the next morning, he discovered to his amazement that he now had twice as many cats as dogs.

"How many dogs and how many cats did Ali have before all these transformations took place?"

28 ☽

AUTHOR'S NOTE: To interrupt my story for a moment, I would like to tell you that one day Ali sold one of his cats, but got very little money for it. Why did he get so little money?

29 ☽ The Wisdom of Haroun El-Rashid

"I liked that one," said the king. "Tell me another."

"Well," she said, "when Ali grew up, he became a devout believer and went with his friend Ahmed on a pilgrimage to Mecca. One day, they stopped at a small village for their midday meal. Ahmed had five loaves of bread and Ali had only three. Just as they were about to eat, a stranger came by and said that he had no food with him but plenty of money, and asked if he might share their meal. The two travelers agreed, and the eight loaves were equally divided among the three. After the meal was over, the stranger thanked them, laid down eight coins of equal value, and departed.

"The problem now arose of how the eight coins should be fairly divided. Ahmed proposed that he should take five coins and Ali should take three, since Ahmed contributed five loaves and Ali three."

"That sounds like a fair division," said the king.

"Well, Ali thought this arrangement unfair; he felt entitled to somewhere between three and four coins, but admitted to not knowing the exact fraction. Since they could not settle this themselves, they took the problem to the wali, but the wali was also unable to settle it.

" 'Take it to the kazi,' the wali suggested. 'He should be able to settle it.'

"Well, they took the problem to the kazi.

" 'Good heavens,' exclaimed the kazi, 'even Ebenezer the Magician couldn't solve this one! It must be settled by the Ruler of the Faithful himself!'

"So Haroun El-Rashid judged the case, surrounded by a

throng eager to hear the verdict. To the astonishment of Ali and Ahmed, as well as everyone else present, the caliph said: 'Let the man who had five loaves take seven of the coins and the man who had three loaves take only one. Case dismissed.'

"How did Haroun get those numbers, seven and one?"

30 ☽ A Sequel

"Some time later," said Scheherazade, "Ali and Ahmed made another pilgrimage, stopped at a village for their meal, and met another stranger who had money but no food, and he, too, asked to share their meal. This time, Ali had three loaves and Ahmed had two, but the loaves were larger than the loaves of the last episode. The five loaves were equally divided among the three. The stranger then laid down ten coins of equal value and departed. How should the coins be divided this time?"

"Very good," said the king, "but I suggest that we now get some sleep. If I grant you another stay of execution, will you tell me some more tomorrow night?"

"With pleasure," replied Scheherazade.

Which brings us to the next night.

Chapter IV

The Thousand-and-Fifth Night, in Which Scheherazade Relates Some Puzzles of Ancient Origin

The Hunting Expedition

"Tonight," said Scheherazade, "I would like to begin with some good old puzzles that come from a land in the West.*

"A certain king was very fond of hunting, and one day took twenty-four of his knights on a hunting expedition. They stayed several nights in one of the king's hunting lodges in the forest. In this house there were nine rooms. The king slept in the central room, and the twenty-four knights, who were to act as his guards, were to be positioned so that there would be nine on each side of the lodge. They were placed there in rooms in the following manner:"

At this point, Scheherazade drew the following diagram:

*Probably England—Au.

3	3	3
3	king	3
3	3	3

"The knights asked if they might meet in the evenings in one another's rooms for games and jousts. To this the king agreed, but only on the condition that there would always be nine knights on each side of the lodge."

31 ☽ The First Night

"On the first night, the king, before retiring, made his rounds of the lodge and counted the number of knights on each side to see that his orders were being obeyed and that none of the knights had gone to the village which was close by. He found that there were indeed nine on each side, and so he went to bed feeling that all was well.

"But his knights had played a little prank on him! Four of them had actually sneaked away to the village. Yet the remaining knights, by a clever rearrangement, had contrived to maintain the full number of nine on each side of the lodge.

"How did they do this?"

32 ☽ The Second Night

"On the second night, instead of any of the knights going to the village, four of the villagers, who were their friends, came to the hunting lodge disguised as knights, which was against the rules. But when the king looked around, he thought all was well, because there were only nine on each side of the lodge.

"How did they manage this?"

33 ☽ The Third Night

"On the third night, eight visitors came, and now there were thirty-two men (other than the king) in the house, but as the king still found nine on each side, he did not notice the new additions.

"How was this arranged?"

34 ☽ The Fourth Night

"The knights had so much fun with all this that on the next night, they received twelve visitors, not eight! Yet these thirty-six men cleverly arranged themselves to fool the king again.

"How did they do this?"

35 ☽ The Fifth Night

"On the fifth and last night, instead of inviting their friends to the lodge, they arranged matters so that six of them could go to the village, and there would still be nine men on each side.

"How did they do this?"

36 ☽ An Ancient Puzzle

"Those were nice," said the king. "Tell me some more."

"Here is an ancient one of Greek origin," said Scheherazade. "It was propounded in 310 A.D. by a man named Metrodorus.

"Demochares has lived one fourth of his life as a boy, one fifth as a youth, one third as a householder, and has spent thirteen years beyond that.

"How old is he?"

37 ☽ Another Ancient One

"Another old puzzle tells of a man who said: 'If I were to give seven coins to each beggar at my door, I would have twenty-four coins left. I lack thirty-two coins of being able to give them nine coins apiece.'

"How many beggars were there and how many coins did the man have?"

38 ☽ The Puzzle of Ahmes

"What is the oldest puzzle you know?" asked the king.

"The oldest one I know," replied Scheherazade, "is contained in an Egyptian papyrus that is several thousand years old.[1] It bears the curious title *Directions for Knowing All Dark Things.*"

"Why *dark* things?" asked the king.

"I really don't know," replied Scheherazade. "Anyway, the

[1]Dating from about 1500 B.C. by the Western calendar.—Au.

author was a priest named Ahmes, and his puzzles were mainly arithmetical. The puzzle I will now tell you is mainly of historical interest, since it is so very simple."

"So what is the puzzle?"

"It is simply this: Find a number that, when added to its seventh part, equals nineteen."

39 ☽ A Hindu Puzzle

"That was indeed simple," said the king. "Give me one less simple."

"All right," said Scheherazade, "here's one that is perhaps less simple. This again is an ancient puzzle, attributed to a famous Hindu mathematician; it goes: 'Beautiful maiden, with beaming eyes, tell me what is the number that, multiplied by 3, then increased by three fourths of the product, then divided by 7, then diminished by one third of the quotient, then multiplied by itself, then diminished by 52, then the square root taken, then increased by 8, then divided by 10, gives the number 2?' "

"Now, really!" said the king. "How do you expect me to figure out anything that complicated?"

"It's really quite simple, if you go about it the right way," said Scheherazade.

How should one go about it?

40 ☽ A Swarm of Bees

"Here is another ancient one by the same Hindu mathematician, and shows the poetic way in which the puzzles of these people are dressed," said Scheherazade. "Translated into our tongue, the puzzle reads: The square root of half the number of bees in a swarm has flown out upon a jasmine bush; eight ninths of the whole swarm has remained behind, one female bee flies about a male that is buzzing within the lotus flower into which he was lured in the night by its sweet scent, but he is now imprisoned in it.

"Tell me the number of bees."

41 ☽ More Bees

"Speaking of bees," said Scheherazade, "I am reminded of another puzzle: In a certain hive, one fifth of the bees flew to a rosebush; one third to some honeysuckle; three times the difference of these numbers alighted on goldenrod, and one flew about a daisy. How many bees were in the hive?"

42 ☽ Two Reports

"Here is another puzzle about bees that combines two ancient ones," said Scheherazade.

"Two boys were in a flower garden one golden afternoon observing the bees that were buzzing about. Both boys were very fond of observing insects, and they each wrote a report of their observations. According to the first report, fourteen of the bees were yellow and the rest were brown. Twelve of the bees were male. Thirteen of the bees were large and the others were small. Four of the yellow ones were large, five of the yellow ones were male, and three of the males were large. There was only one large male yellow bee, and every one of the bees was either large, male, or yellow.

"The second report was quite different. According to this report, half the bees were attracted to clover, a quarter of them were attracted to dandelions, a seventh seemed to favor hyacinths, and the remaining three bees hovered about, without seeming to make up their minds on which flower to alight.

"The question, Auspicious King, is whether either of these reports should be doubted, and if so, which one, or ones? Do the reports agree?"

"Hoomph!" said the king.

What do you make of this?

Chapter V

The Thousand-and-Sixth Night, in Which Scheherazade Gives the King Some Problems in Probability

43 ☽ The Three Chests

Scheherazade began: "Auspicious King, Abdul the Jeweler has in his home three chests of drawers; each chest contains two drawers. In one of the chests, each drawer contains a ruby. In another of the chests, each drawer contains an emerald, and in the third chest, one drawer contains a ruby and the other drawer contains an emerald. Suppose you pick one of the three chests at random and open one of the drawers and find a ruby. What is the probability that the other drawer in the same chest will also contain a ruby?"

"Let me see now," said the king. "Oh yes, the chances are fifty percent."

"Why?" asked Scheherazade.

"Because, once you open a drawer and find a ruby, then the chest with both emeralds is ruled out, and so you have either hit the mixed chest, or the chest with the two rubies, and so the chances are even."

Was the king right?

44 ☽ The Ten Chests

It took Scheherazade quite a bit of time to get the king to accept the correct answer to the last problem, but she finally succeeded.

"I have thought of a related problem," said Scheherazade. "Suppose we have now ten chests instead of three, and each chest has three drawers. Each of the thirty drawers contains either a diamond, an emerald, or a ruby. They are dispersed in the following manner:

1.	D D D		6.	D R R
2.	D D E		7.	E E R
3.	D D R		8.	E R R
4.	D E E		9.	E E E
5.	D E R		10.	R R R

[Of course, *D* stands for diamond; *E* for emerald; and *R* for ruby. So for example, Chest 4 contains one diamond and two emeralds; Chest 7 contains two emeralds and one ruby. There are ten jewels of each of the three types, and they are distributed in all ten possible ways.]

"You open one of the thirty drawers at random and find a diamond. Then you open another drawer of the same chest. What is the probability that it also contains a diamond?"

45 ☽ Two Variants

"As a variant of the problem," said Scheherazade, "suppose you have the same ten chests as before. Again you pick a drawer and find a diamond. And now you are given the option of opening a second drawer from the same chest, or a drawer from one of the other nine chests. If you find a diamond, you may keep it. Would you pick a drawer from the same chest or from a different chest?

"As a second variant, after you have found your first diamond, the last four chests—the ones with no diamonds—are removed, and you are told that these four chests have no diamonds. Thus you know that you are operating with the first six chests. Now you are given the option of opening a second drawer from the same chest, or of opening a drawer from one of the remaining five chests. Which would you do?"

46 ☽ What Are the Odds?

"Here's one," said Scheherazade. "A man has two cats. At least one of them is male. What is the probability that both are male?"

"That's obvious!" said the king.

What is the answer?

47 ☽ What Are the Odds?

"Here's another," said Scheherazade. "A man has two cats—one black and one white. The white cat is male. What is the probability that both are male?"

"Obviously, the same as in the last problem," said the king. "The colors don't make any difference!"

Was the king right?

48 ☽ A Surprising Fact

"Here is a particularly interesting one," said Scheherazade. "Ali and his friend Ahmed agreed to play the following game: Ali would toss a coin. If it came up heads, Ahmed was to pay him two pieces of silver. If it came up tails, Ali was to toss again. If it then came up heads, Ahmed was to pay him four pieces of silver, but if it came up tails, Ali was to toss again. If it then came up heads, Ahmed was to pay eight pieces of silver, but if it came up tails, Ali was to toss again, and so forth. In other words, Ali was to make as many tosses as were necessary for heads to first appear. Then Ahmed was to pay Ali 2^n pieces of silver, where n is the number of tosses. The question, O King, is how much should Ali pay Ahmed in advance to make this a fair game? In other words, what is the expected value of the game?"

"I see no way of telling," said the king. "I would guess something on the order of about a hundred pieces of silver. Am I right?"

The reader not familiar with this problem (which has come down to us under the name of the "Saint Petersburg Paradox") will probably be quite surprised by the solution.

49 ☽ A Controversial Puzzle

"That was a surprise!" said the king, after Scheherazade explained the solution. "Tell me another."

"Here is one of my favorites," said Scheherazade, "and one that provokes much controversy. Suppose I show you three boxes labeled *A*, *B*, and *C*, in one of which there is a prize. The other two are empty. I know which box contains the prize, but you don't. You pick one of the three boxes at random—say, Box A. But before you open it, I open one of the other two boxes that I know to be empty—say, Box B—and show you that it is empty. Then I give you the option of either taking the contents of Box A or trading it for Box C. Is there any probable advantage in your trading?"

"Certainly not," said the king. "Before you showed me the empty box, the chances were one in three that the prize was in Box A. But with the additional information that Box B is empty, the chances are now even that the prize is in Box A or in Box C. So it makes no difference whether I trade or not."

"But I *deliberately* opened a box I knew to be empty," said Scheherazade.

"I don't see what difference that makes," said the king.

"It *does* make a difference," said Scheherazade.

"But it *couldn't!*" said the king.

"Oh yes, it does!" insisted Scheherazade.

Who is right, and why?

The two argued for a long time over this, Scheherazade giving her reasons, and the king giving his. The king finally gave in, whether out of comprehension or sheer exhaustion is not recorded. At any rate, the king fortunately granted Scheherazade another day's stay of execution. This brings us to the next night.

Chapter VI

The Thousand-and-Seventh Night, in Which Scheherazade Relates Some Exploits of Some of Ali Baba's Forty Thieves

50 ☽ Abdul Is Robbed Again

Scheherazade began: "It has reached me, O Auspicious King, that one of Ali Baba's famous forty thieves stole into Abdul's shop and robbed him of some diamonds. Fortunately, they were all recovered, and then it was determined that the thief was either Sabit, Salim, or Shamhir—all of whom were in Ali Baba's famous robber band. At the trial, each of the three accused one of the others, but Shamhir is the only one who lied. Is he necessarily guilty?"

"Not necessarily," replied the king. "An innocent man might lie in order to protect his friend."

Is Shamhir necessarily guilty?

51 ☽ Another Robbery

"Abdul's shop was robbed once more, but the loot was recovered. Again, there were three suspects; their names were Abu, Ibn, and Hasib. At the trial, they made the following statements:

> ABU: I didn't commit the robbery!
> IBN: Hasib certainly didn't!
> HASIB: Yes, I did!

"Later on, two of them confessed to having lied. Who committed the robbery?"

52 ☽ Another Robbery

"Soon after, there was another robbery," said Scheherazade, "and the same suspects—Abu, Ibn, and Hasib—were put on trial. They made the following statements:

> IBN: Hasib never committed the robbery.
> HASIB: That is true.
> ABU: Ibn is innocent.

"Curiously enough, the actual thief told the truth, but they didn't all tell the truth. Which one committed the robbery?"

53 ☽ Another Robbery

"Again, Abu, Ibn, and Hasib were on trial and it was known that one and only one of them was guilty. Abu claimed to be innocent; Ibn agreed that Abu was innocent; and Hasib claimed that he himself was the guilty one. As it turned out, the guilty one lied. Which one was guilty?"

54 ☽ How Many?

"Another sad robbery!" said Scheherazade. "This time one of Ali Baba's men stole a third of Abdul's emeralds. Then, another robber came and stole two thirds of the remaining emeralds. There were then twelve emeralds left. How many emeralds did the first robber find?"

55 ☽ A Hypothetical Theft

"Shortly after," Scheherazade continued, "Abdul replenished his supply of emeralds. He also stocked himself well with diamonds, sapphires, and rubies. For a while, he dealt only in these four kinds of jewels. What is the minimum number of jewels one would have to steal from his shop to be sure of getting at least five jewels of the same kind?"

56 ☽ How Many Bags?

"Here is a simple one," said Scheherazade. "One day, one of Ali Baba's men stole several bags of coins. Each bag contained either sixteen, seventeen, twenty-three, twenty-four, thirty-nine, or forty coins. When he opened all the bags and counted the coins, he found he had a hundred. What bags did he have?"

57 ☽ Another Robbery

"Here is another logic puzzle," said Scheherazade. "One day a valuable sword was stolen. The three suspects were again Ibn, Hasib, and Abu. Ibn claimed that Hasib stole it, and Hasib claimed that Abu stole it. Now, it was not certain that any of these three stole the sword, but it was later determined that no one innocent had lied. Also, the sword was stolen by only one person.

"Can it be determined who stole the sword?"

58 ☽ Another Robbery

"This time a valuable clock was stolen, and it was known that the thief was again either Ibn, Hasib, or Abu. Abu claimed that Hasib was innocent, and Hasib claimed that Ibn was innocent. Ibn's statement was not recorded. Curiously enough, the guilty person told the truth and the two innocent ones both lied. Who stole the clock?"

59 ☽ How Many?

"One night, Abu and Ibn stole some gold coins—all of equal value. 'No fair!' cried Ibn, 'you have three times as many as I!'

" 'All right,' said Abu, 'here are ten more coins.'

" 'Still no fair!' cried Ibn, 'you have twice as many as I!'

"Now the problem, Your Majesty, is how many coins must Abu give to Ibn so that they both have the same number?"

"Just a minute," said the king. "How many coins were stolen altogether?"

"You don't need this information," said Scheherazade.

What is the answer?

60 ☽ A Bit More Greed

"This time, Hasib joined Abu and Ibn in stealing some more gold coins—all of equal value. Again Abu was greedy—the most greedy—since he took three times as many as Ibn, but Ibn wasn't so nice either, since he took twice as many as Hasib. Later, Abu relented and, feeling sorry for Hasib, gave him ten coins. Abu and Hasib then had the same number. How many coins were stolen?"

61 ☽ Honor Among Thieves

"Again the three stole some gold coins—all of equal value. But this time Abu took *all* the coins; he left absolutely nothing for the other two. But they avenged themselves! First, Ibn stole into Abu's dwelling while Abu was asleep, and took five sixteenths of the coins. Later that night, Hasib stole seven elevenths of the coins that remained. When Abu awoke the next morning, he angrily said: 'Now, who has stolen my coins? I have only eight left!'

"How many coins did Ibn and Hasib each get?"

62 ☽ Who Stole What?

"The next robbery involves a logic puzzle," said Scheherazade. "It again concerns Abu, Ibn, and Hasib. One of them stole a horse, one stole a mule, and one stole a camel. The three were finally caught, but—"

"That's a good thing!" said the king.

"But it was not known which thief stole what. There was a trial, and the three made the following statements:

ABU: Ibn stole the horse.
HASIB: Not so; Ibn stole the mule.
IBN: Those are both lies! I didn't steal either.

"As it happened, the one who stole the camel was lying, and the one who stole the horse was telling the truth. Who stole which animal?"

63 ☽ Who Stole What from Whom?

"It is getting late," said the king, "but I think we have time for one more puzzle. Make it a good one!"

"Very well," said Scheherazade. "Let me tell you the most interesting case of all. Three ladies of the sultan's kingdom—Amina, Fatin, and Safie—each owned a precious jewel. One day, three robbers—Abu, Kisra, and Badri—each stole a jewel from one of the ladies, but it was not known who stole what from whom. The case proved extremely baffling, but, fortunately, a wise sage was visiting the country at the time and was able to learn the following facts, which were enough to solve the case:

1. The one who stole the diamond was a bachelor and was the most dangerous of the three thieves.
2. Amina was younger than the lady who owned the emerald.
3. Abu's brother-in-law Kisra, who stole from the eldest of the three ladies, was less dangerous than the one who stole the emerald.
4. The man who stole from Amina was an only child.
5. Abu did not steal from Fatin.

"Who stole what from whom?"

By this time, the king was fast asleep; fortunately, he woke up in a good mood and granted Scheherazade another day's stay of execution. This, then, brings us to the next night.

Chapter VII

In Which Scheherazade Relates Some More Puzzles on This Thousand-and-Eighth Night and Concludes with a Clever Mathematical Observation

"That last puzzle of yesterday was not easy," said the king. "Now I would like some simple ones!"

64 ☽ What Are the Ages?

"Very well, Your Majesty. There are two brothers whose combined age is eleven years. One is ten years older than the other. What are the ages?"

"Oh, come now!" said the king. "I didn't want one *that* simple!"

What are the ages?

65 ☽ How Much?

"All right," said Scheherazade, "a certain animal weighed sixty pounds plus one third of his weight. How much did he weigh?"

"Oh, I've heard that one before," said the king.

For those who haven't, what is the answer?

66 ☽ People Are Not Always So Nice!

"Here's a better one," said Scheherazade. "A group of friends went into an inn to have a meal. The bill amounted to twenty-four coins of equal value, which the men agreed to split equally. But then they discovered that two of the men had slipped away without paying their shares, and so each of the remaining men had to pay one coin more. How many men were originally in the group?"

67 ☽ Ali Baba's Thieves Again

"Here is another one," said Scheherazade. "Ibn once came into a shop and stole one third of the silver coins plus one third of a coin."

"Hold it!" said the king. "How could he steal one third of a coin? He broke off a piece?"

"No, of course not," laughed Scheherazade. "I meant that one third the number of coins he found, plus the number ⅓, is the number of coins that he took, which happened to be a whole number."

"Oh, I see," said the king. "Go on!"

"Well, soon after, Hasib entered the shop and stole one fourth of the remaining coins, plus one fourth of a coin. Soon after, Abu came in and stole one fifth of the remaining coins, plus three fifths of a coin. Finally, another of Ali Baba's thieves came in and stole the remaining 409 coins. How many coins did Ibn find?"

68 ☽ A Simple Logic Puzzle

"How about a logic puzzle," asked the king.

"Very well. Hassan was a good friend of Ali and Ahmed. The following facts are true about them:

1. Either Ali or Ahmed is the oldest of the three.
2. Either Hassan is the oldest or Ali is the youngest.

"Who is the oldest and who is the youngest?"

69 ☽ Which One Is Older?

"Here is another simple logic puzzle," said Scheherazade. "A brother and sister were once asked who was older. 'I am the older,' said the brother. 'I am the younger,' said the sister. It turned out that at least one of the two was lying. Who is older?"

70 ☽ A Trial

"Here is a more interesting one," said Scheherazade. "A man was being tried for having robbed a caravan. Three witnesses came forth and made the following statements:

> FIRST WITNESS: The defendant has committed over a dozen robberies in the past!
> SECOND WITNESS: That is not true!
> THIRD WITNESS: He has certainly committed at least one robbery!

"As it turned out, only one of the witnesses had told the truth.

"Did the defendant rob the caravan or didn't he?"

71 ☽ How Far Is the Shrine?

"Here is an arithmetical one," said Scheherazade. "Ali and his friend Ahmed live equidistant from a certain shrine. They agreed to meet there at a certain time. They both started out at the same time, Ali walking at the rate of five miles an hour and Ahmed walking at four miles an hour. Ali arrived at the shrine seven minutes early, and Ahmed, eight minutes late. How far did each man walk?"

72 ☽ A Hermit's Climb

"Here is another arithmetical one," said Scheherazade. "A hermit started up a mountain trail at eight o'clock in the morning, climbing at the rate of one and a half miles an hour. When he reached the top, he spent twelve hours meditating and resting. Then he went down the same trail at four and a half miles an hour, reaching the bottom at noon the next day. How long was the trail?"

73 ☽ A Clever Student

"And now," said Scheherazade, "I would like to tell you two puzzles that bear an interesting relation to each other and to an important mathematical fact. The first concerns a boy who misbehaved. To punish him, his tutors demanded that he add up all the numbers from one to a thousand."

"That must have taken quite a while!" observed the king.

"The boy was very clever, and gave the answer in a few seconds," said Scheherazade.

"Hmm!" said the king.

How could the boy have done it so fast?

74 ☽ How Many Ways?

"The second," said Scheherazade, "is a problem in probability. Ali thought of a whole number from 1 to 1000 and wrote it down. Then Ahmed thought of a whole number from 1 to 1000 and wrote it down. What is the probability that Ahmed's number is higher than Ali's?"

"Hmm," said the king.

"There are two different ways of solving this," said Scheherazade. "One is shorter than the other, and also more ingenious."

What are the two ways?

75 ☽ Scheherazade's Observation

"The fact that the last problem can be solved in two different ways," said Scheherazade, "yields a new way of proving a famous generalization of the result of the problem before that—the formula $1 + 2 + \ldots + n = \frac{n(n+1)}{2}$."

What could Scheherazade have had in mind?

Chapter VIII

In Which Is Related Scheherazade's Wondrous Account of the Mazdaysians and Aharmanites

On this thousand-and-ninth night, the king said, "Tonight, I am in the mood for more logic puzzles."

"Very well," said Scheherazade. She then began:

76 ☽ The Mazdaysians and Aharmanites

"News has reached me, O Auspicious King, of a curious town in or near Persia in which every inhabitant is either a Mazdaysian or an Aharmanite."

"Oh my goodness, what are *they*?" asked the king.

"The Mazdaysians are worshippers of the Parsi god Ahura Mazda, who is the good god; whereas the Aharmanites worship the evil Parsi god Aharman. The Mazdaysians always tell the truth—they never lie. The Aharmanites never tell the truth—they always lie. All members of any one family are of the same type. Thus, given any pair of brothers, they are either both Mazdaysians or both Aharmanites. Now, I heard a story of two brothers, Bahman and Perviz, who were once asked if they were married. They gave the following replies:

BAHMAN: We are both married.
PERVIZ: I am not married.

"Is Bahman married or not? And what about Perviz?"

77 ☽ Another Version

"According to another version of the story, O Auspicious King, Bahman didn't say that they were both married; instead he said, 'We are either both married or both unmarried.' If that version is correct, then what can be deduced about Bahman and what can be deduced about Perviz?"

78 ☽ A Third Version

"There is still another version, O Auspicious King, that I find the most interesting of the three. In this version, Bahman answered that at least one of the two was married. Then Perviz either claimed that he was married, or he claimed that he wasn't married, but it is not recorded which of those two claims he made. But the person who was interrogating the two brothers was a very wise sage, and could then deduce the marital status of both Bahman and Perviz.

"Is Bahman married or not, and what about Perviz?"

"Now just a minute," said the king. "You haven't given me enough information to solve the problem; you haven't told me what Perviz actually said."

"Yes, I have given you enough information, Your Majesty," replied Scheherazade. "This is what is known as a *metapuzzle.*"

Scheherazade was right! And so what is the solution?

79 ☽ Omar the Magistrate

"I liked those problems," said the king. "Do you know any more problems about these curious people?"

"Yes, indeed, Your Majesty. The wise sage of the last story was named Omar, and he was, of course, a Mazdaysian and was in great demand for his logical skills. For a while, he was the magistrate of the town. Let me tell you some of his encounters.

"Once a man of this town was brought before Omar for having stolen a camel. 'Is it true that you once claimed that you never stole the camel?' asked Omar.

" 'Yes,' replied the defendant.

" 'Did you ever claim that you *did* steal it?' asked Omar.

"The defendant then answered *yes*, or he answered *no*, and after a little thought, Omar was able to decide whether the defendant was innocent or guilty. Which was he?"

"This problem can be solved without knowing what the defendant's second answer was?" asked the king.

"Yes," replied Scheherazade. "This is again a metapuzzle."

What is the solution?

80 ☽ The Town Crier

"A certain town, Your Majesty, has one and only one town crier. Omar was interested in finding the town crier and narrowed the possibilities down to one of three inhabitants. I do not remember their names, so let us call them *A*, *B*, and *C*. They made the following statements:

A: I am not the town crier.
B: The town crier is an Aharmanite.
C: All three of us are Aharmanites.

"Is the town crier a Mazdaysian or an Aharmanite?"

81 ☽ But Which One?

"That's all well and good," said the king, "but that doesn't tell us who is the town crier."

"There is a second part to this story," said Scheherazade. "*A* then made another statement. He said that *C* is an Aharmanite. Does this satisfy Your Majesty?"

82 ☽ Which Ones?

"Another time, Omar interviewed three townsmen—call them *A*, *B*, and *C*. He did not know which were Mazdaysians and which were Aharmanites. Then they made the following statements:

A: Exactly two of us are Mazdaysian.
B: Not so; only one of us is.
C: That is true.

"Which ones are Mazdaysians and which are Aharmanites?"

83 ☽ Which Ones?

"Another time," said Scheherazade, "Omar came across ten inhabitants $A_1, A_2, \ldots A_{10}$ who made the following statements:

A_1: Exactly one of us is an Aharmanite.
A_2: Exactly two of us are Aharmanites.
A_3: Exactly three of us are Aharmanites.
A_4: Exactly four of us are Aharmanites.
A_5: Exactly five of us are Aharmanites.
A_6: Exactly six of us are Aharmanites.
A_7: Exactly seven of us are Aharmanites.
A_8: Exactly eight of us are Aharmanites.
A_9: Exactly nine of us are Aharmanites.
A_{10}: All of us are Aharmanites.

"Which ones are which?"

84 ☽ Innocent or Guilty?

"One day a man of this town was tried by Omar for having stolen an elephant," began Scheherazade.

"That must have been a pretty difficult thing to steal!" said the king.

"I have no idea how it was stolen, but the defendant was actually innocent of the theft. In fact, he made just one statement to the magistrate that clearly proved his innocence but left open the question of whether what he said was true or false. In other words, Omar was convinced of his innocence, but could not tell whether he was Mazdaysian or Aharmanite.

"What statement would work?"

85 ☽ Another Trial

"The next person who was tried for the same crime," said Scheherazade, "made a statement from which Omar could deduce not only the defendant's innocence but also that he was a Mazdaysian.

"What statement could this have been?"

86 ☽ The Next Trial

"The third person tried for stealing the elephant made a statement from which Omar could deduce that the defendant was an Aharmanite but nevertheless innocent of the crime.

"What statement could he have made?"

87 ☽ So Who *Did* Steal the Elephant?

"The next trial," said Scheherazade, "involved *two* defendants, Kushran and Shirin. Here is what happened at the trial:

OMAR (to Kushran): Did you steal the elephant?
KUSHRAN: No, I did not.
OMAR (to Shirin): Do you two worship the same god?

"Shirin then answered—he either answered *yes* or answered *no*. Omar then convicted one of the defendants. Which of the two did he convict, and why?"

"This can be solved without knowing Shirin's answer?" asked the king.

"Yes, it can," said Scheherazade.

What is the solution?

88 ☽ An Intriguing Mystery

"And so," said Scheherazade, "the one who stole the elephant was brought to justice. But finding the *owner* of the elephant proved to be a particularly interesting problem. It was known that the elephant belonged to one of three men, whom we will call *A*, *B*, and *C*. The three made the following statements before the magistrate Omar:

A: The elephant belongs to *C*.
B: I don't own the elephant.
C: At least two of us are Aharmanites.

"From this, Omar could not determine who owned the elephant. 'Come on now,' he said, 'which of you three *really* owns the elephant?'

"*C* answered, naming either *A*, *B*, or himself, and Omar then knew who owned the elephant. Which one was it?"

• •

"Very clever," said the king, "but enough for tonight. How about some more tomorrow night?"

"Fine with me," Scheherazade replied.

And so this brings us to the next night.

Chapter IX

The Thousand-and-Tenth Night, in Which Scheherazade Gives a Further Account of the Mazdaysians and Aharmanites

"I very much like those puzzles about these Mazdaysians and Aharmanites," said the king. "Will you tell me some more?"

"With pleasure," said Scheherazade. "They are among my favorites." And so she began:

89 ☽ At Least One

"Omar the sage met two inhabitants, one of whom made a statement from which Omar could deduce that at least one of them must be an Aharmanite, but there was no way to tell which one—indeed, it could be that both of them are.

"What statement would work?"

90 ☽ At Least One

"Another time," said Scheherazade, "Omar met two inhabitants, one of whom made a statement from which it can be deduced that at least one of them must be a Mazdaysian, but there is no way to tell which. What statement would work?"

91 ☽ Another Time

"Another time," said Scheherazade, "Omar met two inhabitants, one of whom made a statement from which can be deduced that one of them must be a Mazdaysian and the other an Aharmanite, but there is no way to tell which one is which. What statement would work?"

92 ☽ How Many of Them Are Robbers?

"Omar once interrogated two men, Al-Maamun and Ubay, whom he suspected of robbery. Omar first found out that at least one of them must be an Aharmanite, but he had no idea which. Then Al-Maamun said that Ubay was not an Aharmanite robber, and Ubay denied that Al-Maamun was a Mazdaysian who has never committed a robbery.

"How many have committed a robbery?"

93 ☽ Hussein's Robber Band

"Hussein is a fierce robber in a certain town," said Schcherazade. "He is just as dangerous as the noted Ali Baba. To join his robber band, an individual must make a statement that simultaneously accomplishes two things: First, it must convince Hussein that he, the speaker, is an Aharmanite, and second, it must convince Hussein that he, the speaker, has already committed at least one robbery. What statement would do this?"

94 ☽ The Holy Mazdaysians

"To join the Order of the Holy Mazdaysians," said Scheherazade, "one must make a statement that will convince the council that one is a Mazdaysian who has never committed adultery. What statement would do this?"

95 ☽ Case of the Stolen Horse

"In this same town, a horse was stolen by one of the inhabitants. Omar was the presiding magistrate. He asked the defendant, 'Which god is worshipped by the one who stole the horse?' The defendant answered, and Omar then knew whether he was innocent or guilty. Which was he?"

"But what did the defendant answer?" asked the king.

"I needn't tell you that," replied Scheherazade. "In fact, from what I have told you, you can determine what the defendant answered."

What is the solution?

96 ☽ A Solomon-like Case

"Two beautiful ladies, Safie and Zabeide, came before Omar, arguing as to which one was the mother of a certain child.

First, Omar determined that one of them really was the mother, but there was no way yet to tell which one. Then Omar determined that one of the ladies was an Aharmanite and the other was a Mazdaysian, but he still didn't know which was which. Then Omar asked Safie: 'If Zabeide were asked which of you owns the child, what would she answer?' Safie replied, 'Zabeide would reply that the child is hers.'

"Which of the two is really the mother of the child?"

97 ☽ A Logic Puzzle

"Here is a logic puzzle for you," said Scheherazade. "What statement could be made by either a Mazdaysian lady or an Aharmanite man, but could not be made either by a Mazdaysian man or an Aharmanite lady?"

98 ☽ Another Logic Puzzle

"Here is another," said Scheherazade. "What statement could be made by any lady, Mazdaysian or Aharmanite, but could not be made by any man, Mazdaysian or Aharmanite?"

99 ☽ Another Logic Puzzle

"What statement could be made only by a Mazdaysian lady? Neither a Mazdaysian man nor an Aharmanite—lady or man—could make it."

100 ☽ Another Logic Puzzle

"What statement could be made only by an Aharmanite lady?"

101 ☽ Who Is the Traitor?

"Here is an interesting one," said Scheherazade. "It was suspected that there was a traitor in a certain town, but it was not known whether he was a Mazdaysian or an Aharmanite. The three main suspects were named Ayyib, Isa, and Nowas. They were brought before the wise magistrate Omar. Ayyib claimed that Isa was the traitor, and Isa claimed the traitor was Nowas. Then Omar asked Nowas whether all three worshipped the same god. Nowas answered either *yes* or *no*. Omar thought for a while and then said: 'I do not yet have enough evidence to convict any of you, but I do have enough to acquit one of you.' He then pointed to one of the three and said: 'You are clearly not the traitor, so you are free to leave.' The acquitted man then happily left, leaving the two others on trial. Omar then asked one of them: 'Do you two worship the same god?' The one addressed replied, 'Yes,' and Omar then knew who the traitor was.

"Which one was the traitor?"

"Now really!" said the king.

"Yes, really!" replied Scheherazade. "This puzzle can really be solved."

What is the solution?

102 ☽ A Logical Tangle

"Give me one more tonight," said the king.

"Very well," said Scheherazade. "In this case a child of a wealthy sheik was kidnapped and an enormous ransom was demanded and paid. The child was unharmed; still, the kidnapper, or kidnappers, had to be brought to justice. Two men were suspected and tried. It seemed possible at the outset of the trial that neither of the suspects had kidnapped the child, or that just one of them had, or that the two had done it together. No one really knew. The defendants were named Affan and Kurrat. Eight townspeople—*A*, *B*, *C*, *D*, *E*, *F*, *G*, and *H*—came forth as witnesses and made the following statements:

A: Affan worships Mazda.
B: Kurrat worships Aharman.
C: *A* worships Aharman.
D: *B* worships Aharman.
E: *C* and *D* both worship Mazda.
F: *A* and *B* are not both lying.
G: *E* and *F* worship the same god.
H: *G* and I worship the same god, and Affan and Kurrat are not both guilty.

"Out of this logical tangle," said Scheherazade, "the guilt or innocence of each of the defendants can be established."

"Oh my!" said the king.

What is the solution?

Chapter X

In Which Is Related How Scheherazade Entertained the King with Some Special Items on This Thousand-and-Eleventh Night

"That last puzzle you gave me last night was pretty rough!" said the king.

"Well, tonight," said Scheherazade, "I have some special tricks and puzzles for you."

103 ☽ Tricky Numbers

Scheherazade began: "Just think of any three-digit number and write it down, followed by itself. For example, if you are thinking of 294, then write down 294294."

The king thought of 583, and so wrote down 583583.

"Now divide it by 7," said Scheherazade.

The king did this and obtained the number 83369.

"You observe that you have no remainder," said Scheherazade.

"That's right," said the king, "but how did you know? You didn't see what I wrote!"

"Ah!" said Scheherazade. "Now divide the number you have by 11."

The king did this and obtained the number 7579.

"Again, you have no remainder," said Scheherazade.

"Hey, what is this?" exclaimed the king.

"Ah!" said Scheherazade. "Now divide the number you have by 13."

The king did this and obtained the number 583.

"This is a funny coincidence," said the king, "but I happen to have the very number I first thought of."

"That's because you happened to think of a *tricky* number," said Scheherazade.

"What do you mean by a *tricky number*?" asked the king.

"By a tricky number I mean any three-digit number such that if you perform the operations on it that I told you, you will get back the same number."

"That's interesting," said the king. "Are there other tricky numbers?"

"Yes, there are," replied Scheherazade.

"How many are there?"

"Ah, that's the problem I had planned to give you. How many are there?"

"Now, really," said the king, "you don't expect me to go through all the three-digit numbers to test which ones are tricky?"

"Of course not," laughed Scheherazade. "There is a very simple way of telling how many of these 900 numbers are tricky."

How many tricky three-digit numbers are there?

104 ☽ How to Do It

"Here's a little feat that might interest you," said Scheherazade. "With a 7-minute hourglass and an 11-minute hourglass, how would you time the boiling of an egg for 15 minutes?

"There are actually two different ways," continued Scheherazade. "One takes more time, but fewer moves than the other."

Can you find them both?

105 ☽ A Variant

How do you measure 9 minutes with a 4-minute hourglass and a 7-minute hourglass?

106 ☽ Catch the Black King

"Do you play chess?" asked Scheherazade.

"Well, of course!" replied the king.

"Very well. Suppose the black king is on a corner square and a white knight is on the corner square diagonally opposite. No other pieces are on the board. The players alternate moves, and the knight moves first. If the knight can check the king within fifty moves, then White wins; otherwise, Black wins. Which side would you rather play, Black or White?"

107 ☽ A Good Logic Puzzle

"Suppose," said Scheherazade, "that someone put down a copper coin, a silver coin, and a gold coin and asked you to make a statement, with the understanding that if your statement is true, then you will be given one of the three coins, not saying which one, but if your statement is false, then you will be given no coin. What statement could you make that would guarantee that you would get the gold coin?"

"This can really be done?" asked the king.

"Yes," said Scheherazade.

What statement would work?

108 ☽ An Arithmetical Fact

"Let me ask you this," said Scheherazade. "Consider the infinite decimal .999999 . . .—that is, the decimal point followed by an infinite string of 9's. Is this number less than 1?"

"Well, of course!" said the king.

"Less by how much?" asked Scheherazade.

"Hmm, that's hard to say!" said the king.

What would you say?

109 ☽ The Bouncing Ball

It took Scheherazade a long time to get the king to accept the correct answer to the last problem, but she finally succeeded.

"Now, here is an interesting one," said Scheherazade. "Suppose a ball is dropped from a height of 180 feet and on each rebound rises exactly one tenth of the height from which it fell. What is the total distance the ball travels before coming to rest?"

"Hmm," said the king.

What is the answer? (Hint: The previous problem is relevant.)

110 ☽ Dividing the Spoils

"Twenty of Ali Baba's thieves," said Scheherazade, "once captured an enormous plunder consisting of diamonds, emeralds, rubies, sapphires, amethysts, pearls, gold coins, silver coins, ornaments, silks, spices, and other goodies. Then came the question of dividing the loot so that each of the twenty believed he had his fair share. Now, some prefer jewels, some prefer coins, some prefer silks, and so forth. In other words, different members will have different ideas as to what constitutes one twentieth of the loot. So it won't work for one person to simply divide the loot into what *he* considers twenty equal parts and pass the parts around; many would be dissatisfied. However, it *is* possible to make a division such that each person feels satisfied that he has got at least one twentieth of the loot. Of course, there is nothing special about the number twenty; the same could be done with a hundred people, or any number of people. Do you see how to do it with only two people?"

"I don't believe I do," said the king.

"Oh, it's very simple! One person divides the material into what he considers two equal parts and lets the other person choose which part he wants. Then both are bound to be satisfied."

"Oh yes," said the king. "I think I have heard this before. Now, what if there are three people: How is it done?"

"If you figure out how to do it with three," said Scheherazade, "I'm quite sure you will know how to do it with any number."

What is the solution?

AUTHOR'S NOTE: There are actually two quite different solutions to this well-known problem, and many of you readers will know at least one. But the other is of equal interest, and it is well to know both.

111 ☽ A PARADOX

"And now," said Scheherazade, "I have a paradox for you. There are three boxes labeled *A*, *B*, and *C*. One and only one of the three boxes contains a gold coin; the other two are empty. I will prove to you that regardless of which of the three boxes you pick, the probability that it contains the gold coin is one in two."

"That's ridiculous!" said the king. "Since there are three boxes the probability is clearly one in three."

"Of course it's ridiculous," said Scheherazade, "and that's what makes it a paradox. I will give you a proof that the probability is one in two, and your problem is to find the error in the proof—since the proof must obviously contain an error."

"All right," said the king.

"Let's suppose you pick Box A. Now, the coin is with equal probability in any of the boxes, so if Box B should be empty, then the chances are fifty-fifty that the coin is in Box A."

"Right," said the king.

"Also, if Box C is empty, then again the chances are fifty-fifty that the coin is in Box A."

"That's right," said the king.

"But at least one of the boxes, B or C, must be empty, and whichever one is empty, the chances are fifty-fifty that the coin is in Box A. Therefore, the chances are fifty-fifty, period!"

"Oh, my!" said the king.

What is the solution to the paradox?

Chapter XI

Scheherazade's Puzzles and Metapuzzles of the Thousand-and-Twelfth Night

"Tonight," said Scheherazade, "I have some unusual metapuzzles for you. But first, I would like to give you two little warm-up puzzles."

112 ☽ Scrambled Labels

"Here is a little problem in probability," said Scheherazade. "As in a problem I gave you several nights ago, we have three chests, each of which contains two drawers. Each drawer contains an emerald or a ruby. Again, one chest contains two emeralds, one chest contains two rubies, and one chest contains one emerald and one ruby. And now, we have three labels to be attached, one to each chest. One label reads *'E E,'* meaning both emeralds; one reads *'R R,'* meaning both rubies; and one reads *'E R,'* meaning one emerald and one ruby. Now, suppose the labels are put on randomly, so that some might be right and some wrong. What is the probability that only one is wrong?"

113 ☽ More Scrambled Labels

"Continuing with the set-up of the last problem," said Scheherazade, "suppose that in fact all three labels are wrong, and you are told to open one drawer at a time until you are able to determine the contents of all three chests. What is the smallest number of drawers you have to open in order to do this?"

114 ☽ A Scrambled Label Metapuzzle

"Here is a more interesting one," said Scheherazade. "This time, four men were given a test. In front of each man was a chest of three drawers. Each drawer contained either an emerald or a ruby. One chest contained three emeralds, one contained three rubies, one contained two emeralds and one ruby, and the remaining chest contained one emerald and two rubies. The four labels *'E E E,' 'R R R,' 'E E R,'* and *'E R R,'* respectively, had been put on the four chests and the four men were each told that his label was wrong. No man could see any label but his own. Each was to open two drawers of his chest and try to determine what jewel lay in the third drawer. The first man opened two drawers and said: 'I've found two emeralds and I know what the remaining jewel is.' The second man then opened two of his drawers and said: 'I've found one emerald and one ruby, and I know what my remaining jewel is.' Then the third man opened two of his drawers and said: 'I've found two rubies.' The experimenter who was conducting the test said: 'That's all well and good, but do you know what jewel your third drawer contains?' The man then answered *yes*, or *no*, but we are not told what he answered. Now, if you were told what the third man answered, you would be able to figure out what each of the four chests contained and which label was on each chest."

"Whee," said the king, "this problem seems to be hanging in midair! You say that *if* I were told I could get the solution, but since I am not told, then I presume I can't get the solution. Am I right?"

"Ah, but you *can* get the solution, since you now have the additional information that *if* you were told, then you could get the solution. This additional information is enough to get the solution."

What is the solution?

"That was quite an elaborate problem!" said the king, after Scheherazade had explained the solution.

"Yes," said Scheherazade, "that was a metapuzzle."

"You have used that word several times now," said the king. "Just what is a metapuzzle?"

"Oh," said Scheherazade, "it's the kind of puzzle in which you are not given complete information but can solve the puzzle only on the basis of knowing that someone who had more information was or was not able to solve it."

"That's very interesting!" said the king. "Do you know any more metapuzzles?"

"Oh yes," replied Scheherazade. "Here's a nice one."

115 ☽ What Are Their Ages?

"Iskandar was an extremely intelligent person who once asked his friend Kamar the ages in years of his three children. The following conversation ensued:

KAMAR: The product of their ages is thirty-six.
ISKANDAR: That doesn't tell me their ages.
KAMAR: Well, by coincidence, the sum of their ages is your own age.
ISKANDAR (after several minutes of thought): I still don't have enough information.
KAMAR: Well, if this will help, my son is more than a year older than both his sisters.
ISKANDAR: Oh good! Now I know their ages.

"What are their ages?"

116 ☽ What Type Is Bulikaya?

"I liked that one," said the king. "Give me another metapuzzle."

"Very well," said Scheherazade. "I'll give you a metapuzzle about the Mazdaysians and Aharmanites. As you recall, the Mazdaysians worship the good god Mazda, and always tell the truth; whereas the Aharmanites worship the evil god Aharman and always lie. Now, one of the inhabitants of this town was named Bulikaya.

"He was a rather mysterious individual and no one seemed

to know whether he was a Mazdaysian or an Aharmanite. Well, at one point it became quite important to find out, and so the town council asked Omar to investigate the case. Omar came across Bulikaya with his friend Ayn Zar. Now, just because they are friends doesn't mean that they worship the same god."

"How could two friends not worship the same god?" asked the king.

"These people are very tolerant of other people's religions," replied Scheherazade. "Anyhow, Omar asked Ayn Zar: 'Is either of you a Mazdaysian?' Ayn Zar answered, but Omar could not tell what Bulikaya was. Then Omar asked Bulikaya: 'Did Ayn Zar answer truthfully?' Bulikaya answered, and then Omar knew whether he was a Mazdaysian or an Aharmanite. Which was he?"

"Now, this is very baffling," said the king. "Is it really possible to solve this without being told what *either* one answered?"

"Yes, it is," replied Scheherazade.

What is the solution?

Chapter XII

The Thousand-and-Thirteenth Night, in Which Scheherazade Relates the Story of Al-Khizr

"Tonight," said Scheherazade, "I will give you some puzzles of a completely different nature than any I've given you so far."

Scheherazade then related the following story and puzzles.

117 ☽ The First Test

A certain prince Al-Khizr was in love with the sultan's daughter and asked for her hand in marriage.

"My daughter is very choosy," said the sultan, "and wants to marry only someone who shows extraordinary intelligence. So if you want to marry her, you must first pass eight tests."

"What are the tests?" asked the suitor.

"Well, for the first test, you have to write down a number that will be sent to the princess. She will then send back a number to you. If she sends back the very same number that you have sent her, then she will allow you to take the second test. But if her number is different from yours, then you are out."

"Now, how can I possibly know what number to write?" asked the suitor. "How can I guess what number the princess has in mind?"

"Oh, she doesn't have a number in mind," said the sultan. "The number she sends back is dependent on the number you send. The number you send *completely determines* what number she will send back. And if you send the right number, then she will send back the same number."

"Then how can I guess the right number?" asked the suitor.

"It's not a matter of *guessing*," said the sultan. "You must *deduce* the correct number from the rules I am about to give you. For any numbers *x* and *y*, by *xy* I mean not *x times y* but *x followed by y*, both numbers, of course, written in base ten Arabic notation. For example, if *x* is 5079 and *y* is 863, then by *xy* I mean 5079863. Now here are the rules:

> *Rule 1*: For any number *x*, if you write her *1x2*, then she will send you back the number *x*. For example, if you write 13542, she will write back 354.
>
> *Rule 2*: For any number *x*, the *repeat* of *x* means *xx*. For example, the repeat of 692 is 692692. And now, the second rule is that if *x* brings back *y*, then *3x* will bring back the repeat of *y*. For example, since 15432 brings back 543, then 315432 will bring back 543543. From which it further follows that if you send her 3315432, you will get

back 543543543543 (since 315432 brings back 543543).

Rule 3: The *reverse* of a number means the number written backwards. For example, the reverse of 62985 if 58926. The third rule is that if *x* brings back *y*, then *4x* brings back the reverse of *y*. For example, since 172962 brings back 7296, then 4172962 brings back 6927. Thus, if you send her the number 4172962, you will get back 6927. Or, combining Rules 1, 2, and 3, since 316982 brings back 698698 (by Rules 1 and 2), then 4316982 brings back 896896.

Rule 4 (The Erasure Rule): If *x* brings back *y*, and if *y* contains at least two digits, then *5x* brings back the result of erasing the first digit of *y*. For example, since 13472 brings back 347, then 513472 brings back 47.

Rule 5 (The Addition Rule): If *x* brings back *y*, then *6x* brings back *1y* and *7x* brings back *2y*. For example, since 15832 brings back 583, then 615832 brings back 1583, and 715832 brings back 2583.

"Those are the rules," said the sultan, "and from them can be deduced a number *x* that will bring back the very number *x*. There are actually an infinite number of solutions, but any single one will suffice for passing the first test."

"Are there any *meanings* to these numbers?" asked the suitor.

"Ah, that is the princess' secret, but fortunately you don't have to know the meaning in order to pass the first test."

118 ☽ The Second Test

For the second test, the suitor had to send the princess a number, x, such that she would send back the *repeat* of x—the number xx. What x would work?

119 ☽ The Third Test

For the third test, the suitor had to send the princess a number x such that she would send back the *reverse* of x. What number x would work? An extra bonus would be given if the number x contains no more than twelve digits.

What x would work?

120 ☽ The Fourth Test

For this test, the suitor had to send a number x such that the princess would send back the number x with its last digit erased. What x would work?

121 ☽ The Fifth Test

For this test, the suitor had to send a number x such that the princess would send back a *different* number y, which the suitor was to send back to the princess, and she would (hopefully) send back the first number x.

What number x would work?

122 ☽ The Sixth Test

The suitor now had to send a number x, get back a number y, return y to the princess, and get back the *reverse* of the original number x.

What number x would work?

123 ☽ The Seventh Test

For this test, the suitor was to send a number x, get back a number y, return y to the princess, and get back the number x with the first and last digits interchanged.

What number x would work?

124 ☽ The Eighth Test

For the final test, the suitor was to send a number x; the princess would then send back a number y; the suitor was then to send her the *reverse* of this y; the princess would then send back a number of the form zz (a number z repeated); the suitor was then to break zz in half (so to speak) and send her back z. The princess would then (hopefully) send back the original number x.

What number x would work?

Chapter XIII

The Grand Question!

We now come to the most remarkable part of the entire story!

"Those last puzzles made my head ache, and you have kept me up all night!" exclaimed the king. "I think I've had enough puzzles for a lifetime! Enough! Enough, I say! I think it's high time for your execution. The dawn is breaking, and today's as good a day as any."

"So be it," said Scheherazade, "but you wouldn't deny a condemned lady her last request, would you?"

"That depends on the nature of the request," replied the king. "What do you have in mind?"

"I will ask you a question," said Scheherazade, "a question answerable by *yes* or *no*. All I ask is that you answer *yes* or *no*, and that you promise to answer truthfully."

"I always answer questions truthfully," replied the king.

"Then you promise?"

"Yes, of course!"

Scheherazade then framed a question so cleverly that the king, to keep his word, had to spare her life! Now, there are three different versions as to what actually happened. My secret source gives all three versions and honestly states that it is unknown which of the three is correct. Here are the three versions:

125 ☽ The First Version

According to this version, the question was of such a nature that the king, to keep his word, had no option but to answer *yes* and spare her life.

What question would work?

126 ☽ The Second Version

According to this version, the king had to answer *no* and spare her life.

What question would work?

127 ☽ The Third Version

According to this version, the king had the option of saying either *yes* or *no*, but in either case, he had to spare her life.

What question would work?

Epilogue: These days, one seldom ends stories happily; one rarely writes that the couple lived happily ever after. However, if I am to be truthful and give an accurate account of my secret source, then I have to inform you that the king and Scheherazade *did* live happily ever after—very happily, in fact. They had many beautiful and bright children who grew up to become makers of all sorts of puzzles, which have come down to us through the ages.

Book Two

FROM SCHEHERAZADE TO MODERN LOGIC

Chapter XIV

Coercive Logic

The question Scheherazade had asked the king had an almost magical quality in that it forced him to do something he wouldn't otherwise have done—namely, to spare her life. My son-in-law, Dr. Jack Kotik, came up with a perfect name for this kind of logic—*coercive logic*—a name that I happily adopt. This seems to me to be a promising subject, and so, in this and the next two chapters, I will start the ball rolling with more problems in coercive logic. (From now on, solutions to problems are given at the end of each chapter.)

1 ☽ Another Tale of Scheherazade

As stated at the end of Book I, Scheherazade and the king formed a happy union. A couple of years after the last related episode, the king was again in the mood for logic puzzles.

"Here is one you might like," said Scheherazade. "A certain sultan had two lovely daughters and said to a visiting prince, 'You seem like a personable young man, and I like you, so here is what I am going to do: You are to make a statement. If the statement is true, then I will give you my younger daughter in marriage, but if the statement is false, then you may not marry her.'

"Now, it so happened that the prince had his heart set on the *elder* daughter, not the younger! He then cleverly made a statement that forced the sultan to give him the elder daughter. What statement would do this?"

2 ☽ A Variant

"There is another version," said Scheherazade. "According to this one, the prince was greedy and wanted to marry *both* daughters. What statement could the prince make to force the sultan to give both daughters?"

3 ☽ A Third Version

"There is yet another version," said Scheherazade. "In this one, the sultan had *three* daughters, not two. He said to the prince, 'If you make a statement that is true, then I will give you at least one of my daughters in marriage—maybe two, or maybe all three. But if your statement is false, then you may not marry any of my daughters.'

"Now, it so happened that the prince was in love with the youngest daughter and the eldest one, but had no desire to marry the middle daughter. What statement could he make that would force the sultan to give him just the two daughters of his choice?"

4 ☽ A Wily Student

Coming now to modern times, a logic student was once on a date and said to the young lady: "I'd like to ask you a little favor. I will make a statement. All I ask is that if the statement is true, you give me a photograph of yourself. Will you do that for me?" The lady assented. "But also," continued the man, "if my statement is false, then I want you to promise *not* to give me your photograph. Agreed?" The lady agreed.

The man then cleverly made a statement such that the lady, after thinking of it for a while, realized (to her secret amusement) that in order to keep her word, she would have to give him not her photograph but a kiss!

What statement would work?

5 ☽ How to Win Both

There is another statement the man could have made so that the lady would have to give him both her photograph and a kiss.

What statement would do this?

6 ☽ A Variant

There is yet another statement the man could have made so that the lady would have the option of giving him her photograph or not, but in either case she would have to give him a kiss.

What statement would work?

7 ☽ Forcing You to Tell the Truth

Speaking of coercion, there is a yes/no question I could ask you that, in answering, it would be *logically impossible* for you to lie! (I'm assuming that you answer either *yes* or *no*.) You are *forced* to tell the truth!

What question would work?

8 ☽ Forcing You to Lie

Then, of course, there is another yes/no question that would force you to answer falsely (assuming you answer either *yes* or *no*). What question would work?

I wish to interrupt these coercive logic puzzles for a moment to tell you a funny thought that came to mind while reading Plato.

In one of Plato's dialogues, after Socrates criticizes the sophist Protagoras for taking money from his students and not teaching them anything of real value, Protagoras responds: "At the end of my course, if the student feels that he has not learned anything of value, then I give him his money back."

Now, let us suppose that at the end of the course, a student comes to Protagoras and demands his money back. Protagoras then asks, "Can you give me a good argument why I should give you your money back?" The student then gives an excellent argument, upon which Protagoras says, "You see the dialectical skill I have taught you?"

Then a second student comes to Protagoras demanding his money back. Again, Protagoras asks, "Can you give me a good argument why I should give you your money back?" The student thinks a while and says, "No." Protagoras then says, "All right, here's your money back."

9 ☽ Which Would You Choose?

Coming back to coercive logic, here is a variant of an old problem of mine which affords a good illustration and which, together with the next problem, is good preparation for Problem 11.

Arthur and Robert each make you the following offers: Arthur asks you to make a statement and promises to pay you exactly ten dollars if the statement is true, but if the statement is false, then he will pay you either more or less than ten dollars, not saying which, but not exactly ten.

Robert, on the other hand, asks you to make a statement and offers to pay you twenty dollars if the statement is true, and nothing if the statement is false. Which of the two offers would you accept?

10 ☽ Winning the Greater Good

For many years now, in teaching logic to liberal arts students, I have done the following: I place a penny and a quarter on the table and ask the student to make a statement. If the statement is true, then I agree to give the student either the penny or the quarter, not saying which one, but if the statement is false, then I will give neither coin. What statement could the student make that would force me to give the quarter?

11 ☽ I Could Have Been a Victim!

As I said, I had been doing the above for many years, until one day I suddenly realized to my horror that all that time I had left myself wide open to losing a million dollars! Yes, there is a statement the student could have made that would have forced me to hand him or her a million dollars (assuming I kept my word). What statement would accomplish this?

Solutions

1 • Another Tale of Scheherazade A statement that would work is: You will not give me either daughter in marriage. If the statement were true, then the sultan would have to give him the younger daughter, as agreed, but that would falsify the statement. Therefore, the statement can't be true, which means that it is not the case that he will give neither daughter, and so he will give at least one of the two daughters. But he can't give the younger one, since the statement is not true, hence he has to give the elder one.

2 • A Variant A statement that works is: You will not give me the younger daughter in marriage unless you also give me the elder daughter.

The only way the statement can be false is if the sultan gives the prince the younger daughter without giving the elder daughter as well, but this is not possible, since the sultan said that he wouldn't give the younger daughter if the statement were false. Therefore, the statement can't be false; it must be true. Since it is true, the sultan must give the younger daughter, as agreed, but then he also has to give the elder daughter.

3 • A THIRD VERSION A statement that works is: You will either give me none of your daughters in marriage, or just the youngest and eldest one. The sultan can't give none of the daughters, as that would make the statement true. And a true statement gets at least one daughter. Thus, the sultan must give at least one daughter, but the only way he can do that is if the statement is true. Thus it is true that he will give either none of the daughters or just the youngest and eldest ones. But he can't give none of the daughters, since the statement is true, and therefore he must give just the youngest and eldest ones.

4 • A WILY STUDENT The man said, "You will not give me either your photograph or a kiss." The underlying logic is really the same as that of Problem 1 (just substitute "photograph" for "younger daughter" and "a kiss" for "elder daughter").

5 • HOW TO WIN BOTH A statement that works is: You will not give me your photograph unless you also give me a kiss. The underlying logic is the same as that of Problem 2.

6 • A VARIANT Here the logic is different from that of any of the preceding problems in this chapter.

A statement that works is: You will either give me both your photograph and a kiss or neither one. Suppose the lady gives him her photograph. If she didn't also give him a kiss, the statement would be false, but she said she wouldn't give the photograph for a false statement. And so if she gives the photograph, she must also give him a kiss.

Now, suppose she doesn't give him the photograph. If she also didn't give him a kiss, then his statement would be true, but she can't withhold the photograph if the statement is true! And so, if she doesn't give her photograph, then again she must give him a kiss. And so she has the option of giving her photograph or not, but in either case she must give him a kiss.

7 • Forcing You to Tell the Truth A question that works is: Will you answer *yes* to this question? If you answer *yes*, you are affirming that *yes* is your answer, and you are right! If you answer *no*, then you are denying that *yes* is your answer, and again you are right! And so whether you answer *yes* or *no*, you are right, hence you cannot answer falsely even if you want to.

8 • Forcing You to Lie For this, a question that works is: Will you answer *no* to this question?

9 • Which Would You Choose? Most people would choose Robert's offer, thereby guaranteeing the possibility of getting twenty dollars, whereas (so they think) the most that can be guaranteed with Arthur's offer is ten dollars (by making a true statement). However, Arthur's offer is really vastly preferable, for all you need say is: "You will give me neither exactly ten dollars nor a million dollars." If the statement

is true, then Arthur must pay you ten dollars as agreed, but doing so would falsify the fact that he neither pays you ten or a million. Thus it is contradictory that the statement is true, hence it must be false. Since it is *false* that Arthur will give you *neither*, it must be the case that he will give you *either*. But he cannot give you exactly ten dollars for a false statement, hence he has no alternative but to give you a million dollars! (If that's not coercion, I'd like to know what is!)

10 • WINNING THE GREATER GOOD All the student need say is: You will not give me the penny. If the statement were false, that would mean that I *would* give him the penny, but I can't give a penny for a false statement. Therefore, the statement must be true, which means that I won't give him the penny. Yet I must give the student one of the two coins for making a true statement, and since it is not the penny, I must give the quarter.

11 • I COULD HAVE BEEN A VICTIM! To force me to give a million dollars, all the student need say is: You will give me neither the penny, nor the quarter, nor a million dollars. By an analysis similar to that of Problem 9, the statement must be false and I must pay a million dollars.

Chapter XV

Right- and Left-Handed Coercion

In my previous puzzle book, *Satan, Cantor, and Infinity*, I gave a group of puzzles about a country in which every inhabitant was either right-handed or left-handed; none of the inhabitants was ambidextrous. Moreover, whatever a right-handed person wrote with his right hand was true, and whatever he wrote with his left hand was false. The left-handed people were the opposite: Whatever they wrote with their left hand was true and whatever they wrote with their right hand was false. In other words, whatever a person wrote with his stronger hand was true, and whatever he wrote with his weaker hand was false. As typical puzzles of this group, we have:

1. What statement could be written only by a left-handed person, and using either hand?
2. What statement could be written only by a left-handed person using his right hand?

3. What statement could be written only by a left-handed person using his left hand?

Answers:

1. I wrote this with my left hand.
2. I am a right-handed person who wrote this with my left hand.
3. Either I am left-handed or I wrote this with my left hand.

Now, for those who like unusually difficult puzzles, in this chapter I will combine the above scenario with the coercive logic of the last chapter, and so our set-up will be this: You write a letter to an inhabitant of this strange country, and the letter consists of a single question answerable by *yes* or *no*. If it is possible to answer the question at all, he or she will do so, either truthfully with the stronger hand, or falsely with the weaker hand.

1 ☽

What question could you write that would force both a right-handed and a left-handed person to answer with the left hand, but he or she would have the option of answering either *yes* or *no*?

2 ☽

What question could be answered only by a left-handed person using his or her right hand and answering *no*?

3 ☽

What question would force the answerer to lie?

4 ☽

What question could you write to determine whether the answerer is right-handed or left-handed?

5 ☽

What question could you write to determine whether the answerer is male or female?

6 ☽

What question could be answered either *yes* or *no* by any female, using either hand, but by no male?

7 ☽

What question would force the answerer to answer *no*, and the person could be either left-handed or right-handed and could use either hand?

8 ☽

What question could you ask that no inhabitant of this country could possibly answer?

9 ☽

What question could you ask that could be answered only by a right-handed male with his left hand, writing *no*?

Solutions

1 • A question that works is: Are you either a left-handed person who will answer *yes* to this question, or a right-handed person who will answer *no*?

You are asking whether one of the following two alternatives holds:

1. You are left-handed and will answer *yes.*
2. You are right-handed and will answer *no.*

There are now 4 possible cases: He truthfully answers *yes*; he falsely answers *yes*; he truthfully answers *no*; he falsely answers *no.*

Case 1: He truthfully answers yes. Then one of the above two alternatives really does hold. It cannot be Alternative 2 since he answers *yes* so it must be Alternative 1. Thus he is left-handed, and since he answered truthfully, he must have used his left hand.

Case 2: He falsely answers yes. Then neither of the two alternatives holds—particularly the first. Then he must be right-handed (because if he were left-handed, Alternative 1 *would* hold [since he *did* answer *yes*]). So he is right-handed and answered falsely, so he must have used his left hand.

Case 3: He truthfully answers no. Then again, neither Alternative 1 nor Alternative 2 holds. In particular, the second doesn't hold, but since he answered *no*, then he can't be right-handed—he must be left-handed. And he answered truthfully, so he used his left hand.

Case 4: He falsely answers no. Then one of the two alternatives *does* hold. It cannot be the first, so it must be the second. So he is right-handed and answered falsely, so he used his left hand.

In all 4 possible cases, the answerer must use his left hand.

2 • The question you can ask is whether one of the following three alternatives holds:

1. You will answer *yes* with your weaker hand.
2. You will answer *no* with your stronger hand.
3. You are left-handed and will answer *no* with your right hand.

Suppose you get the answer *yes*. If it is truthful, then one of the three alternatives really does hold, but it can't be the first or the third (since a truthful answer can't be given with the weaker hand), nor the second (since the answer was *yes*), so we have a contradiction. Therefore, a truthful *yes* answer is not possible. But a false *yes* answer is also impossible, since it would imply on the one hand that none of the alternatives really does hold, yet a false *yes* answer means a *yes* answer with the weaker hand, which is Alternative 1. What about a true *no* answer? This would mean that none of the three alternatives holds, yet the second *must* then hold! So a true *no* answer is out. This leaves a false *no* answer, which means one of the three alternatives really does hold, which obviously must be Alternative 3.

3 • Will you answer *no* to this question? (I leave it to the reader to prove that this works.)

4 • Will you answer this with your right hand? (If you get the answer *yes*, he is right-handed, and if the answer is *no*, he is left-handed.)

5 • Are you either a male who will answer this with your stronger hand, or a female who will answer this with your weaker hand? (A male will answer *yes*, either truthfully or falsely, whereas a female will answer *no*, again either truthfully or falsely.)

6 • A question that works is: Are you either a female who will answer *yes* with your stronger hand or *no* with your weaker hand, or a male who will answer either *yes* with your weaker hand or *no* with your stronger hand?

I leave the proof to the reader.

7 • Will you answer this with your weaker hand?

8 • Will you either answer *no* with your stronger hand, or *yes* with your weaker hand?

9 • Ask whether one of the following three alternatives holds:

1. You will answer *yes* with your weaker hand.
2. You will answer *no* with your stronger hand.
3. You are a right-handed male who will answer this with your left hand.

The proof that this question works is very similar to that in the solution for Problem 2.

Chapter XVI

The Ultimate in Coercive Logic

1 ☽ The Ultimate in Coercive Logic

Suppose I offer you a million dollars to answer a yes/no question truthfully, would you accept the offer? If so, you shouldn't, for I would then ask: Will you either answer *no* to this question or pay me two million dollars? The only way you can answer truthfully is by answering *yes* and then paying me two million dollars (by the same argument as the analysis of Scheherazade's question in Problem 125, Book One).

But now, suppose I offer you a million dollars to answer either *yes* or *no*, and you have the option of answering either truthfully or falsely—you may either tell the truth or lie at your pleasure. Surely now you are completely safe, are you not? How can mere *words* coerce you into paying me money when you don't even have to be truthful in your answer? Why can't you simply answer *yes* or *no* at random and then refuse to pay me anything? How much safer can you be?

Well, are any of you game? You had better not be! (See the solution!!)

2 ☽ A Variant

(To be read after the solution of the last problem.)

There is a question I could ask that is different from the one given in the solution to the last problem and is such that the only way you can fulfill the conditions of the offer (in other words, that you answer *yes* or *no*, and that your answer be true or false) is by either falsely answering *yes*, or truly answering *no*, and in either case you must pay me two million dollars. What question would work?

3 ☽ Another Variant

There is another question I could ask such that if you don't pay me two million dollars, then neither your *yes* answer nor your *no* answer could be either true or false, but if you do pay me, then your *yes* answer could be either true or false without contradiction, and there is no way to tell which, and your *no* answer could also be either true or false and there is no way to tell which! Thus, you could either truthfully answer *yes* and pay me, or falsely answer *yes* and pay me, or truthfully answer *no* and pay me, or falsely answer *no* and pay me, but if you didn't pay me, then you couldn't answer either *yes* or *no*, either truthfully or falsely.

What question would work?

Solutions

1 • The Ultimate in Coercive Logic The whole idea is to frame a question in such a manner that if you don't pay me two million dollars, then, whether you answer *yes* or *no*, the answer you give cannot be either true or false without involving a contradiction, and must then be paradoxical! Thus, the only way you can avoid a paradoxical answer is by paying me two million dollars.

Remember, my offer was not that you merely answer *yes* or *no*, but that your answer be also either true or false! Of course, you can answer at random and not pay me the two million dollars, but if I ask the right question, you cannot answer *truthfully* or *falsely* without paying me the two million dollars. What question would work? Before giving the solution, it will be instructive to first consider a slightly simpler problem: What question could I ask you such that it would be impossible for you to answer *yes* or *no* either truthfully or falsely—that is, either answer, *yes* or *no*, would be paradoxical. (The reader might like to try solving this before reading ahead.)

Well, one such question is: Will you either truthfully answer *no* to this question or falsely answer *yes*? What I am asking is whether either of the following two alternatives holds.

1. You will truthfully answer *no* to the question.
2. You will falsely answer *yes*.

Suppose you answer *yes*. Could this answer be true? If it is, then either Alternative 1 or Alternative 2 really does hold (as you have affirmed), but the first doesn't (since you answered *yes*), hence the second does, and thus you have *falsely* answered *yes*, contrary to the assumption that your answer was true. Thus it is logically impossible for your *yes* answer to be true. Could your *yes* answer be false? If so, Alternative 2 does hold, hence *either* the first or second holds, which means that your *yes* answer was true after all! And so it is equally contradictory to assume that your *yes* answer could be false! Thus, if you answer *yes*, your answer cannot be either true or false, but only paradoxical. And so you cannot either truthfully or falsely answer *yes*.

Can you truthfully answer *no*? If you do, then Alternative 1 holds, hence either of the two alternatives holds, so your *no* answer was then false! Again, a contradiction, and so you cannot truthfully answer *no*. What about a false *no* answer? Well, if you answered *no* falsely, then you didn't truthfully answer *no*, so the first alternative doesn't hold, and since you didn't answer *yes*, then the second doesn't hold, hence neither the first nor the second holds, which means that your *no* answer was true after all! Therefore it is equally contradictory to assume that your *no* answer could be false. And so neither a *yes* answer nor a *no* answer could be either true or false; either answer must be paradoxical!

Now, the following modification of the above question can be answered nonparadoxically, but only if you pay me two million dollars!

> Will you either truthfully answer no to this question, or falsely answer yes, or pay me two million dollars?

I am asking whether one of the following alternatives holds:

3. You will truthfully answer *no* to this question.
4. You will falsely answer *yes*.
5. You will pay me two million dollars.

Suppose you answer *yes*. You are thus affirming that either the third, fourth, or fifth alternative holds. If your *yes* answer were false, then on the one hand, none of the alternatives would hold (since you wrongly asserted that one of them did), but on the other hand, the fourth would hold, which is a contradiction. Therefore, your *yes* answer cannot be false. So, it must be true (assuming that you lived up to the terms of the offer and answered either truthfully or falsely), which means that one of these three alternatives *does* hold. Well, obviously Alternative 3 doesn't hold (since you answered *yes*), and the fourth doesn't hold (since you *truthfully* answered *yes*), hence the only way your *yes* can be *true* is if the fifth holds, which means that you must pay me two million dollars. (If you didn't pay me the two million dollars, then your *yes* answer couldn't be either true or false without involving a contradiction!)

Now, suppose you answer *no*. If your answer is true, then on the one hand none of the alternatives holds (as you correctly claimed by answering *no*), but on the other hand, the third does hold, which is a contradiction. Therefore, your *no* answer cannot be true, hence it is false (assuming you have lived up to the conditions of the offer), which means that one of the three alternatives *does* hold. The only one that can hold in this case is Alternative 5, and so again, you must pay me two million dollars.

In summary, you have the option of truthfully answering *yes* and paying me, or falsely answering *no* and paying me. (If you don't pay me, you cannot answer either truthfully or falsely, but only paradoxically.)

2 • A VARIANT A question that works is the following:

> Will you either truthfully answer *no* and not pay me two million dollars, or falsely answer *yes* and not pay me two million dollars?

I am asking whether one of the following two alternatives holds:

1. You will truthfully answer *no* and not pay me two million dollars.
2. You will falsely answer *yes* and not pay me two million dollars.

We shall assume that your answer is either true or false.

Suppose you answer *yes*. If your answer were true, we would have a contradiction (on the one hand, either alternative holds, but then also neither could hold), so your answer is false. Hence, neither Alternative 1 nor Alternative 2 holds, and so, in particular, the second doesn't hold, so you must pay me two million dollars (because if you didn't, the second *would* hold!).

Suppose you answer *no*. If your answer were false, then on the one hand either the first or second alternative would hold, yet neither one could hold, which is contradictory. So your answer must be true, hence neither alternative holds, and then, since the first doesn't hold, you must pay me (for if you didn't, the first *would* hold).

And so with this question, your only options are either to falsely answer *yes*, or truthfully answer *no* (unlike the previous question, in which you must either truthfully answer *yes* or falsely answer *no*), and in either case you must pay me.

3 • ANOTHER VARIANT A question that works is: Will you either truthfully answer *yes* to the question and pay me two million dollars, or falsely answer *no* and not pay me, or truthfully answer *no* and not pay me, or falsely answer *no* and pay me?

I leave the proof to the reader.

Chapter XVII

Variable Liars

We shall now leave coercive logic and go on to other things.

1 ☽

Many puzzles have been written about the Island of Knights and Knaves, on which knights always tell the truth, knaves always lie, and every inhabitant is either a knight or a knave. But now we shall visit a particularly interesting island on which each day, each inhabitant either lies the entire day or tells the truth the entire day. An inhabitant might lie on some days and tell the truth on others, but during any one day, the inhabitant's behavior is constant.

Take Jal, for example: He lies only on Mondays and tells the truth the other six days of the week. One day he said, "Today is Monday and I am married." Was it really Monday? Is he really married?

2 ☽

According to another version of the above story, Jal didn't say, "Today is Monday *and* I am married." What he said was, "Either today is Monday *or* I am married." (*Or* means at least one, or possibly both.)[1]

If this version is correct, is Jal married or not, and is it really Monday?

3 ☽

There is still a third version, somewhat like the first version but subtly different. Instead of making the single statement "Today is Monday and I am married," Jal first said, "Today is Monday." Then, sometime later that same day, he said, "I am married."

What should one make of this version?

4 ☽

What statement could Jal make on Thursdays but not on any other day?

[1]This is the way that *or* is used in logic and computer science. In ordinary English, the word *or* is sometimes used *exclusively*, meaning exactly one, and sometimes *inclusively*, meaning at least one. For example, if I say that tomorrow I will marry either Ethel *or* Gertrude, I obviously mean one and only one. But if a college catalogue states that for admission an applicant must have either a year of mathematics *or* a year of a foreign language, the college certainly won't exclude you if you happen to have both!

In this book, *or* will be used to mean at least one, and possibly both.

5 ☽

It so happens that Jal has a brother Tak who lies on Thursdays and on no other days. One day, one of the two brothers said, "Tomorrow is Tuesday." Exactly one week later he said: "I will lie tomorrow." On what day of the week was this?

6 ☽

According to another version of the story, after one brother said, "Tomorrow is Tuesday," it was the *other* brother who, one week later, said, "I will lie tomorrow." If this version is correct, then on what day of the week was it?

7 ☽

One day the elder brother said, "I am Jal." Then the other brother said, "I am Tak." Who is older, Jal or Tak?

8 ☽

On this island, corresponding to every inhabitant A is an inhabitant A' who tells the truth on those and only on those days on which A lies. In other words, on any day that A lies, A' tells the truth, and on any day that A tells the truth, A' lies. The behavior of A' is always *opposite* to that of A.

A second feature of this island is that for any inhabitants A and B, there is an inhabitant C who tells the truth on all days on which A and B both tell the truth, and on no other days (C

lies on any day on which at least one of *A* and *B* lies). There is a rumor that no one on this island tells the truth on all days. Is this rumor true or not?

Solutions

1 • If it were any day other than Monday, Jal could never lie and say that it was Monday and he is married. Therefore, the statement must have been made on Monday. It then follows that the statement must be false. Now, if he were married, the statement would be true (since the day really was Monday), but the statement isn't true, so he is not married.

Thus the statement was made on Monday, but Jal is not married, so his statement as a whole was false.

2 • Suppose the second version is the correct one. Then, if the statement had been made on Monday, it would be true that *either* the day is Monday *or* Jal is married, but Jal couldn't make such a true statement on Monday. Therefore, the day of the week is some day *other* than Monday, and hence true. Since it is true that *either* the day is Monday *or* he is married, but the day isn't Monday, hence Jal is married (again, if the second version is correct).

Note that the conclusions of the second version are the opposite of those of the first! In the first, the day is Monday and Jal isn't married. In the second, the day isn't Monday and Jal *is* married!

3 • As to the third version, it is simply false! On no day could Jal make the single statement that it is Monday, since on Monday he would never truthfully say that, and on any other

day he would never falsely say that. And so the third version is simply wrong.

Although Jal could never make the single statement "Today is Monday," he could make the compound statement "Today is Monday *and* I am married" (assuming he is not really married).

This illustrates an interesting and important difference between the logic of lying and the logic of truth-telling: If a truthful person asserts of two propositions that they are both true, then he can assert each one separately. But with a liar, this is not so if one of the propositions is true and the other false.

4 • The statement "Today is Thursday" won't quite work, because although Jal could say that on Thursday, he could also say it on Monday. A statement that does work is "Today is either Monday or Thursday." Jal couldn't say that on a Monday, because the statement would be true. And if the statement were made on any day other than Monday or Thursday, the statement would be false, but Jal is truthful on such days. Therefore, the only possibility is that Jal truthfully made the statement on a Thursday.

5 • To say that tomorrow is Tuesday is tantamount to saying that today is Monday, which, as we have seen, Jal couldn't have said. Therefore, the brother in question must be Tak. Then Tak could have said that only on a Monday (truthfully) or on a Thursday (falsely), and so the day was either a Monday or a Thursday. Now, on the same day of the week, Tak said that he would lie tomorrow. The only days he could have said that were Wednesday (truthfully) or Thursday (falsely). Hence the day was not Monday, so it was Thursday (and Tak lied both times).

6 • Let us suppose that this version is correct. Then again it was Tak who said that tomorrow is Tuesday, and again the day on which it was said was either Monday or Thursday. But now it was Jal who, on the same day of the week, said, "I will lie tomorrow." Well, the only days Jal could have said that is either Sunday (truthfully) or Monday (falsely). And so the day must be Monday.

7 • If one statement is true, then so is the other, because the two statements are either both true or both false. They can't both be false, since the two brothers never lie on the same day. Therefore, both statements are true, hence Jal is the older brother.

8 • Take any inhabitant A. Then there is an inhabitant A' whose lying habits are always opposite to those of A. Then, by the second condition (taking A' for B) there is an inhabitant C who tells the truth on those and only those days on which A and A' both tell the truth. But A and A' never tell the truth on the same day, hence C never tells the truth! Thus, C lies on all days. But then there is an inhabitant C' who behaves opposite to C on all days, hence C' tells the truth on all days! Thus the rumor is false.

Chapter XVIII

Chinese or Japanese?

A certain computer scientist designed a remarkable series of machines that gave correct answers to any *yes/no* questions put to them. The machines would answer by flashing either a green light or a red light—one of the colors meant *yes*, the other *no*. The machines were manufactured in China and Japan, but were unfortunately not manufactured uniformly, in that the machines made in one of these two countries meant *yes* by green and *no* by red, but the machines of the other country meant *no* by green and *yes* by red. To add to the confusion, it was not known whether it was the Chinese or the Japanese machines whose green meant *yes*.

1 ☽

Now, suppose you come into possession of one of these machines and wish to find out whether it was made in China or Japan. You are allowed to ask the machine only one *yes/no* question. What question would you ask?

2 ☽

Suppose that instead of wanting to know whether your machine is from China or Japan, you wanted to know whether it was the Chinese or the Japanese machines that meant *yes* by green. What single *yes/no* question could you ask the machine to determine this?

3 ☽

Suppose that all you wanted to know about your machine is which of its two colors meant *yes* and which meant *no*. What *yes/no* question would accomplish this?

4 ☽

What *yes/no* question is such that the machine, regardless of whether it is Chinese or Japanese, is bound to answer by flashing red?

5 ☽

What *yes/no* question is such that the Chinese machines will have to answer by flashing red, and the Japanese machines by flashing green?

6 ☽

There is a *yes/no* question you could ask the machines that would make it impossible for it to answer with either red or green! What question would work?

7 ☽

There is a *yes/no* question that the machines whose green means *yes* could answer with either red or green, but the machines whose green means *no* couldn't answer at all. What question would work?

8 ☽

The first batch of machines made in these two countries are quite rare and are valuable collector's items. What *yes/no* question could you ask your machine to determine whether or not it is a collector's item?

9 ☽

Suppose you go into a store to purchase one of these machines. "I have only three left," says the proprietor, placing them on the counter, "and unfortunately one of them was faultily constructed and flashes red and green completely at random, and I don't know which of the machines are the good ones!" Well, it is possible by asking just one *yes/no* question to one of the machines to find one you know to be good. What question would work?

Solutions

1 • A question that works is: Do the Chinese machines mean *yes* by green?

Suppose you get green for your answer. There are two possibilities—either the green means *yes* or it means *no*. Suppose it means *yes*. Then, since the machine is accurate, it really is the case that the Chinese machines mean *yes* by green, and since your machine flashed green and meant *yes*, it must be Chinese. On the other hand, suppose the machine means *no* by flashing green. Then it is not the case that the Chinese machines mean *yes* by flashing green; they mean *no*. But then your machine flashed green and meant *no*, so again your machine must be Chinese. This proves that regardless of whether your machine's green means *yes* or *no*, if it flashes green to your question, it must be Chinese. A similar analysis, left to the reader, reveals that a red answer would indicate that the machine is Japanese.

2 • A question that works is: Were you made in China?

Supposing you get green for an answer. If green means *yes*, then the machine really was made in China, and hence the Chinese machines mean *yes* by green. On the other hand, if green means *no*, then your machine was not made in China, it was made in Japan; hence the Japanese machines are the ones that mean *no* by green, and so again, it is the Chinese machines that mean *yes* by green. Thus, if you get green for an answer, then regardless of what green means, it is the Chinese machines that mean *yes* by green. A similar analysis reveals that if you get red as an answer to the question, then it must be the Japanese machines that mean *yes* by green.

3 • The solution to this is so obvious that it tends to be overlooked! Just ask whether two plus two is four. Whatever color is flashed must mean *yes*.

4 • Ask: Does red mean *yes*? If the machine's red means *yes*, it will flash red. If its red means *no*, then *no* is the correct answer to your question, hence the machine will again flash red. In either case, the machine will flash red.

5 • This is essentially Problem 1 all over again: Just ask the machine whether the Chinese machines mean *yes* by red. The Chinese machines will answer with red and the Japanese with green.

6 • Ask: Does the color that you answer to this question mean *no*? If the color that the machine answers does mean *no*, then the machine cannot correctly answer with that color. On the other hand, if the answering color means *yes*, then it is also wrong! Thus the machine cannot correctly answer the question.

7 • Ask: Will you answer this question with green?

Suppose the machine's green means *no*. Then both green and red would be wrong answers to the question. On the other hand, if the machine's green means *yes*, then both green and red would be correct answers to the question.

8 • A question that works is: Do you flash green when asked whether you are a collector's item?

Suppose the machine flashed green. If green means *yes*, then the machine really does flash green when asked whether it is a collector's item, and since green means *yes*, the machine really is a collector's item. On the other hand, if green means *no*, then the machine doesn't flash green when asked if it is a collector's item; it flashes red, but since red then means *yes*, the machine is again a collector's item. Thus, whatever green means, a green answer indicates that the machine is a collector's item. A similar analysis reveals that a red answer would indicate that the machine is not a collector's item.

Another question that works is: Is it the case that either you are a collector's item and your green means *yes*, or that you are not a collector's item and your green means *no*?

9 • Call the machines *A*, *B*, and *C*. Ask *A*: When asked whether *B* is good, do you flash green? Suppose *A* flashes green. Then if *A* itself is good, *B* must be good (by the same argument as in the solution of the last problem), and if *A* is not good, then of course *B* is good. Thus, regardless of whether *A* is good or not, a green answer means that you are safe buying *B*. By a similar analysis, a red answer means that you can safely buy *C*.

Chapter XIX

Oron and Seth

In a certain solar system, the sun has only two planets—Oron, inhabited by the Oronians, and Seth, inhabited by the Sethians. Both races are highly intelligent and able to travel from one planet to the other. The trouble is that whenever the inhabitants of either planet land on the other one, they become totally disoriented and all their beliefs are wrong! When they are back on their home planet, they are perfectly oriented again and all their beliefs are then correct.

1 ☽

Here is a simple starter: An inhabitant from one of these two planets once believed that he was an Oronian who was then on Seth. Was he Oronian or Sethian, and what planet was he on at the time?

2 ☽

What proposition could be believed by inhabitants of both planets, regardless of whether or not they are on their home planet at the time?

3 ☽

It is reported that an inhabitant of one of the planets once said, "I believe that I am not now on my home planet."

Is that really possible?

4 ☽

Intermarriages between Oronians and Sethians are quite common. At any given time, a man and his wife might be on the same planet or on different planets. On one occasion, a man was speaking by radiophone to his wife, who was then on the other planet. "After all," said the husband, "we are both Oronians."

"That is not true!" said his wife.

Which of the two was on Oron at the time?

5 ☽

Og and Belinda are a mixed couple—one is Oronian and the other is Sethian. At one point, Og believed that he and his wife were on different planets, but at the same instant, Belinda believed that they were on the same planet. Which one was right?

6 ☽

"Are you Oronian or Sethian?" Tak was once asked.

"I am Oronian," he replied.

"And your wife?"

"Oh, my wife and I are both Sethian."

"Hm!" said the interrogator.

Is Tak Oronian or Sethian? And what about his wife? And on which planet did this strange conversation occur?

7 ☽

A certain female celebrity of one of these two planets was once interviewed and asked, "Tell me about your parents. Where are they from?"

"My father is Oronian and my mother is Sethian," she replied.

A few weeks later, she was on the other planet and was interviewed again and asked, "Are your parents from the same planet, or are they a mixed couple?"

"A mixed couple," she replied.

Is her father Oronian or Sethian? And what about her mother?

8 ☽ A Metapuzzle

One day a crime was committed on one of the two planets. It was not known whether the criminal was Sethian or Oronian, but it was known that his name was Murdoch. A trial took place, and the judge and jury and court recorder were all on their home planet. There was only one defendant, but it was not known whether he was Oronian or Sethian, or whether he was Murdoch. The judge then asked him, "Are you Sethian or Oronian, and are you Murdoch?" The records of the trial are unfortunately not clear as to the response, which was either "I am Sethian, but I am not Murdoch," or it was "I am Oronian, but I am not Murdoch," and the court recorder could not for sure remember which.

At any rate, the judge then knew whether the defendant was Murdoch.

Was he?

Solutions

1 • Obviously, it must have been a Sethian then on Oron.

2 • One such proposition is: I am now on my home planet.

3 • On neither planet can an inhabitant of that planet or the other planet believe that he is now not on his home planet, but he could wrongly believe that he believes it (since he doesn't believe it). So it *is* possible that an inhabitant once said that!

4 • If one of them was Sethian and the other Oronian, they couldn't have disagreed (because either they would both be on their home planets, hence both correct in their beliefs, or both off their home planets, hence both wrong in their beliefs), but they did disagree, hence they are either both Oronians or both Sethians. If the former, then the husband was right, hence he would be on Oron (since he is then Oronian). If the latter, then the wife was right, and being Sethian, she must be on Seth. So in either case, the husband was on Oron and the wife on Seth.

5 • Since they are a mixed couple and disagreed, then they must be on the same planet, hence Belinda was right.

6 • Obviously, Tak was off his rocker at the time, hence off his home planet, and his answers were both wrong. Therefore, he is not Oronian as he claimed, he is Sethian. Then, since it is false that he and his wife are both Sethian, she must be Oronian. Finally, since he was not on his home planet, he was on Oron.

7 • Since she made the two statements on different planets, one of them must be right and the other wrong. If the first statement was right, the second statement would also have to be right, which cannot be. Hence, the first statement must have been wrong and the second right. Since the second was right, her parents really were a mixed couple. But the first statement was wrong, and so her father must be Sethian and her mother Oronian.

8 • A METAPUZZLE The solution does not depend on which planet the trial took place on, so let us momentarily

assume that it was on Seth. Suppose the defendant said that he was Sethian but not Murdoch. If he is Sethian, then his statement was correct and he is really not Murdoch, but if he is Oronian, his statement was wrong regardless of whether he is or is not Murdoch, and there is no way of telling which. Therefore, if he said that he was Sethian but not Murdoch, the judge couldn't have known if he was dealing with Murdoch. On the other hand, if he said that he was Oronian but not Murdoch, the judge would know that he must be Oronian (since no Sethian on Seth could say that), hence his statement must be false, and hence he must be Murdoch. Since the judge did know, then it must be that the defendant claimed to be Oronian but not Murdoch, and the judge then knew that he was Oronian and Murdoch.

By a symmetric argument, if the trial took place on Oron, then the defendant would have to say that he was Sethian but not Murdoch (otherwise the judge couldn't have known whether he was Murdoch), and the judge then knew that he was Sethian and Murdoch.

In short, we cannot know on which planet the trial took place, nor what the defendant said, but we do know that, in effect, the defendant said that he was from the other planet but not Murdoch. In reality, he *was* from the other planet, and he was Murdoch.

Chapter XX

Which Personality?

Jack and John are twin brothers and both have dual personalities. Jack in his normal state always tells the truth, but in his altered state always lies. John is the opposite; in his normal state, John always lies, but in his altered state, he always tells the truth. The two are indistinguishable in appearance; their mother is about the only person who can tell them apart. Several interesting problems arise:

1 ☽

Suppose you meet one of the brothers one day and wish to know whether he is Jack or John. You may ask him only one question answerable by *yes* or *no*, and the question must be simple, not compound—that is, it must not contain logical connectives such as *and*, *or*, *not*, *if/then*. There is a perfectly simple, straightforward question that will do the job. What question will work?

2 ☽

Another time you meet one of the brothers and wish to know not whether he is Jack or John but whether he is in his normal or altered state. What simple yes/no question would determine this?

3 ☽

Suppose that instead of wanting to know what state a twin is in, you wish to know whether both brothers are in the same state or not. What simple yes/no question would you ask?

4 ☽

If you meet one of the two, it is obvious how you can tell whether he is in a truthful state. Just ask him a question such as whether two plus two equals four. But suppose you want to know not whether *he* is now truthful, but whether his brother is now truthful. What simple yes/no question will accomplish this?

5 ☽

On a certain day, one of the brothers was in the same state all day long, and made only one statement that day: "Tomorrow

I will be in my altered state." The next day, he was in the same state all day long, and made only one statement that day: "Yesterday I was not truthful."

Is he Jack or John?

6 ☽

One day a medical doctor visited one of the brothers accompanied by the twins' mother. The doctor didn't know whether it was Jack or John but knew which state he was in. The mother knew which one it was but didn't know which state he was in. Then the man said, "I am John and am now in my altered state." The doctor still couldn't tell whether it was Jack or John, and the mother still couldn't tell which state he was in.

Was it Jack or John, and which state was he in?

Solutions

1 • You have merely to ask: Are you in your normal state? Jack in his normal state will truthfully say *yes*, and in his altered state will falsely say *yes*. John in his normal state will falsely say *no*, and in his altered state will truthfully say *no*. And so, if the man answers *yes*, he must be Jack, and if he answers *no*, he must be John.

2 • You have merely to ask: Are you Jack? Both brothers in their normal state will say *yes* (Jack truthfully and John falsely), and both in their altered state will say *no* (Jack falsely and John truthfully). Thus, a *yes* answer indicates that the

man is in his normal state, and a *no* answer indicates that he is not.

There is a pretty symmetry between this and the last problem: To find out whether he is Jack, you ask him whether he is in his normal state, whereas to find out if he is in his normal state, you ask him if he is Jack.

3 • Ask him: Is your brother now truthful? If he truthfully answers *yes*, then the other brother really is now truthful, hence the two are in different states. If he falsely answers *yes*, then the other brother is now not truthful, so again the two are in different states (both lying). Thus, a *yes* answer indicates that the two are in different states.

A similar analysis reveals that a *no* answer indicates that they are in the same state (one truthful and the other not).

4 • Ask: Are you now in the same state as your brother?

Suppose he answers *yes*. If he is now truthful, then his brother really is in the same state, hence he is not truthful. If the speaker is lying, then his brother is now in a different state, hence he is also lying. Thus, a *yes* answer reveals that the brother is now not truthful. By a similar analysis, a *no* answer indicates that the brother is now truthful.

The same pretty symmetry exists between this problem and the last, as between the first two problems: Of the two questions Is your brother now truthful? and Are you in the same state as your brother?, to find out the answer to either question, you ask the other one.

5 • Suppose that on the second day he was truthful. Then what he said on the first day was really false, which means that on the second day he was actually in his normal state, and being truthful on that day, so he must be Jack.

On the other hand, suppose he lied on the second day. Then he was actually truthful on the first day, so on the second day he was really in his altered state. In this case, he is again Jack (since he lied during his altered state). Thus, in either case, he must be Jack.

6 • One thing is clear: Jack in his normal state could never claim to be John in the altered state, so if the doctor had known that the man was in his normal state, then he would have known that the man couldn't possibly be Jack, hence must be John. But the doctor didn't know, and therefore the man must have been in his altered state. On the other hand, if the mother had known that it was her son Jack, she would have known that he was in an altered state because he made an obviously false statement. But she didn't know, and so she must have known that it was John. And so the answer is that the man was John in his altered state (and spoke truthfully).

Chapter XXI

Oh No!

There is a still more curious pair of identical twins named Edward and Edwin, who are similarly indistinguishable in appearance. One day shortly after they were grown, a strange disease struck them both and changed their lives forever. Henceforth, each one was in one of three psychological states—State 1, State 2, or State 3—that alternated in a constant cyclical pattern—1, 2, 3, 1, 2, 3, . . . and so on. Curiously enough, at any given time, both brothers were always in the same state—both were either in State 1, State 2, or State 3. But there was a crucial difference. Edward always lied when he was in State 1, but told the truth in the other two states. On the other hand, Edwin lied when in State 2, but told the truth when in either of the other two states.

1 ☽

One day I met the two brothers out for a walk. One of them said, "I am Edward." The other then said, "I am Edwin." There was no change of state between the two statements.

Which one was Edward: The one who spoke first, or the one who spoke second?

2 ☽

On another occasion, I met the two, and one said, "I'm Edward." The other (who hadn't changed his state) then said, "If that's true, then I am Edwin." Later, I found out that they were not in State 3 at the time.

Which one was Edward?

3 ☽

One day, one of the brothers said, "My last state was a lying state." Then the other brother (who hadn't changed his state) said the same thing.

Which state were they then in?

4 ☽

One day, Edward made the following two statements:

1. My last state was a lying state.
2. My next state will be a lying state.

Edward didn't change his state between these two statements. In what state was he in?

5 ☽

One day, one of the brothers said, "I am Edward and am now in State 1."

Who was he?

6 ☽

Another time, one of the brothers said, "Either I am Edward *or* I am in State 2." (Remember, *or* means at least one and maybe both.)

Who was he?

7 ☽

Another time, one of the brothers said, "Either I am Edward or I am now in a lying state."

Who was he, and was he truthful at the time?

8 ☽

One day, one of the brothers was asked, "Are you either Edwin in State 2 or Edward not in State 1?" From his answer, is it possible to deduce what state he is in? From his answer, can one deduce whether he is Edward or Edwin?

9 ☽

Suppose you meet one of the brothers one day and wish to find out which state he is in. This cannot be done with certainty using only one yes/no question, but it can be done with two.

What two yes/no questions would work?

10 ☽

Suppose you meet both brothers one day. There is a single yes/no question such that if you ask it of each of them in turn, you can then know what state they are in (assuming no change of state has occurred between the two answers).

What question would work?

11 ☽

One of the brothers once lost a silver watch that was found by a neighbor who came to the house to return it. When asked to whom the watch belonged, one brother said, "Edward owns the watch." And the other brother said, "I am Edward." There was no change of state between the two statements, and the brothers were not in State 3 at the time.

Was it the first or the second twin who owned the watch?

12 ☽

Someone once reported that he had visited the two brothers and one of them had said, "We are now in State 1." And the other one then said, "What he just said was true."

Does this report hold water? (NOTE: This answer is a bit tricky!)

13 ☽

A logician once met one of the brothers and asked him, "Are you now in State 1?" The brother answered (*yes* or *no*) and the logician then deduced whether he was speaking to Edward or Edwin.

Which brother was it?

14 ☽

What yes/no question is such that in each of the three states, both brothers will always give opposite answers?

15 ☽

One of the brothers is married and the other is not. I once met one of the brothers and asked him whether he was married. He replied, "The married one of us is now in a lying state."

What is the probability that the one I spoke to is married?

16 ☽

On another occasion I met one of the brothers (either the same one as before or not, but I didn't know which), and again I asked whether he was married. He replied, "The married one is now in a truthful state."

What is the probability that this one is married? (I am assuming that the three states are equally probable and that the chances are even that I spoke to Edward or Edwin.)

17 ☽ Another Metapuzzle

One of the two brothers was a spy and the other was not. There was a trial to determine which one was the spy. The court first ascertained which one was Edward and which was Edwin, but it was not known which state they were in at the time. The judge first asked Edward, "Are you the spy?" Edward replied, "Yes." Then the judge asked Edwin, "Are *you* the spy?" Edwin answered—(either *yes* or *no*)—and the judge then knew which one to convict. (He correctly assumed that there was no change of state between the two responses.)

Which brother is the spy?

Solutions

1 • If the first one is really Edward, then the second one is really Edwin, and if the first one is not really Edward, then the second one is not really Edwin, and so either they both told the truth or they both lied. However, there is no state in which

they both lie, hence they both told the truth, and so the first one is Edward.

2 • The second one obviously made a true statement, and since they were not in State 3, the other one was lying, and hence was really Edwin.

3 • The only states in which Edward could say that are State 1 and State 2. The only states in which Edwin could say that are State 2 and State 3. Thus, they were in State 2.

4 • Edward's first statement could only be made in State 1 or State 2. His second statement could only be made in State 1 or State 3. So he was in State 1.

5 • Edward couldn't say that in any state, because in State 1, it could be true that he is Edward in State 1, but Edward in State 1 doesn't make true statements, and in State 2, Edward would never lie and claim to be in State 1, hence the speaker must be Edwin. Clearly, he lied, so he was in State 2.

6 • Edwin couldn't say that in any state, because in State 1 or State 3 (his truthful states), he is neither Edward nor in State 2; whereas in his lying state it is true that he is *either* Edward *or* in State 2 (since he is then in State 2), and so he wouldn't then make that true assertion. Therefore, the speaker is Edward (in State 2 or State 3; there is no way to tell which).

7 • If he were in a lying state, then it would be true that *either* he is Edward *or* he is in a lying state, but a truthful statement cannot be made in a lying state. Therefore, he is in a

truthful state. Thus his statement was true, and so either he is Edward or he is in a lying state. But since he is not in a lying state, he must be Edward.

8 • You cannot tell what state he is in, but you can tell who he is.

Suppose he answers *yes*. If he is in a truthful state, then he really is either Edwin in State 2 or Edward not in State 1. But he then can't be Edwin in State 2 (in which he lies); hence he must be Edward, but not in State 1. On the other hand, if he lied, then, contrary to what he said, he is neither Edwin in State 2 nor Edward not in State 1; hence he is either Edwin not in State 2 (and thus in a truthful state) or Edwin in State 1, but he can't be Edwin not in State 2, since he lied; hence he must be Edward in State 1. This proves that if he answers *yes*, he must be Edward (maybe in State 1 or maybe not).

Now, suppose he answers *no*. If his answer is truthful, then he is neither Edwin in State 2 nor Edward not in State 1; hence he is either Edwin not in State 2 or Edward in State 1. But he can't be Edward in State 1, since he told the truth; so he must be Edwin (but not in State 2). On the other hand, if he lied, then he is either Edwin in State 2 or Edward not in State 1, but the latter alternative is not possible (since Edward not in State 1 doesn't lie), so he must then be Edwin in State 2. Thus, if he answers *no*, he must be Edwin.

In summary, if he answers *yes*, he is Edward, and if he answers *no*, he is Edwin.

9 • You can first find out whether he is in State 1 by asking: Are you Edwin who is now in either State 1 or State 2? Here are the answers you will get:

Edward in State 1 - Yes	Edwin in State 1 - Yes
Edward in State 2 - No	Edwin in State 2 - No
Edward in State 3 - No	Edwin in State 3 - No

Therefore, if you get the answer *yes*, you will know that he is in State 1, and you then do not need to ask another question. If you get the answer *no*, then you will know that he is in State 2 or State 3, but you won't know which. To find out whether he is in State 2, you ask: Are you Edward who is now in either State 1 or State 2? By a similar analysis, you can see that if he answers *yes*, he is in State 2, and if he answers *no*, he isn't, and is then in State 3.

10 • A very simple question that works is: Are you Edward? Here are the possible answers:

Edward in State 1 - No	Edwin in State 1 - No
Edward in State 2 - Yes	Edwin in State 2 - Yes
Edward in State 3 - Yes	Edwin in State 3 - No

Therefore, if you get the answers *no*, *no*, they must be in State 1. If you get *yes*, *yes*, they are in State 2. If you get one *yes* and one *no*, then they are in State 3 (in which case you will know, furthermore, that Edward is the one who answered *yes*).

11 • Since they were not in State 3, one of them told the truth and the other lied. Suppose the first one was truthful, then Edward really owns the watch, but since the second one lied, Edward is the first one, hence it is the first one who owns the watch. Now, suppose, on the other hand, that the first one lied, then Edward doesn't really own the watch, and since the

second one was then truthful, he is really Edward; so he doesn't own the watch. So again, it is the first one who owns the watch. Thus in either case, the one who spoke first is the owner of the watch.

12 • Yes, the situation is possible, but only if there was a change of state between the two statements!

To begin with, Edward could not have made the first statement, for if he were in State 1, they would both be in State 1, and his statement would then be true, which is not possible for Edward in State 1. On the other hand, if Edward were not in State 1, he would be truthful and not claim that they were in State 1. Thus, it was Edwin who made the first statement. He was then either in State 1 and told the truth, or in State 2 and lied. Suppose the former. Then Edward was also then in State 1. If he hadn't changed his state between the two statements, he would never have agreed with Edwin's true statements. Hence, he must have changed to State 2, and then truthfully agreed that Edwin was right when he said what he did. Thus, if Edwin was in State 1 when he spoke, then Edward changed his state before he spoke. Now, on the other hand, what if Edwin was in State 2 when he spoke? Then he lied and Edward was also in State 2; regardless of whether he changed to State 3, he would never have made the false statement that Edwin spoke the truth. So this case is out!

In summary, Edwin made the first statement and was in State 1 at the time, and then they changed to State 2 and Edward correctly corroborated Edwin's statement.

13 • Edward would answer *no* in all three states. Edwin would answer *yes* in States 1 and 2, and would answer *no* in

State 3. Therefore, if the answerer said *no*, he could be either Edward or Edwin, but if he answered *yes*, he could only be Edwin. Since the logician *did* know who he was from his answer, then the answerer must have been Edwin and answered *yes* (and must have also been in State 3). This is a *metapuzzle*.

14 • Many solutions are possible: One such is the question: Are you Edward and now in State 3? In States 1, 2, and 3, Edward would answer, respectively, *yes*, *no*, *yes*, whereas Edwin would answer, respectively, *no*, *yes*, *no*.

15 • The probability is zero! If he were in a truthful state at the time, then the married brother was really in a lying state at the time, hence was not the speaker. On the other hand, if the speaker were in a lying state at the time, then the married brother was not really in a lying state at the time, hence again could not have been the speaker.

16 • In this case, the chances are 5 out of 6 that the speaker is married. Here's why: The speaker is with equal probability in State 1, State 2, or State 3; hence, the probability that he is either in State 1 or State 2 is 2 out of 3 or, what is the same thing, 4 out of 6. Now, if he is in either State 1 or State 2, then he is definitely the married one, because he and his brother are then opposite with respect to truth-telling. And so if the speaker is truthful, then the married one is really in a truthful state; hence, it can't be his brother who is in a lying state, so it must be the speaker. On the other hand, if the speaker is in a lying state, then his statement was false, and the married one is not really in a truthful state—he's in a lying state—so he must

be the speaker (since his brother is then in a truthful state). This proves that if the speaker is in either State 1 or State 2, then he is definitely the married one.

Now, we consider the case that he is in State 3 (the probability of which is 1 in 3). Well, if he is in State 3, then both brothers are truthful at the time, and also the speaker's statement was true; hence the married brother really is in a truthful state, but could then be either the speaker or his brother with equal probability. Thus, if the speaker is in State 3, the probability that he is the married brother is 1 in 2—or 50–50. Thus, the probability that the speaker is *both* in State 3 *and* married is 1 in 2 × 1 in 3, which is 1 in 6.

In summary, the probabilities that the speaker is in State 1 and married, State 2 and married, State 3 and married are, respectively, 1 in 3, 1 in 3, and 1 in 6; hence, the probability that he is married is 1 in 3 + 1 in 3 + 1 in 6 = 2 in 6 + 2 in 6 + 1 in 6 = 5 in 6.

17 • ANOTHER METAPUZZLE Suppose Edwin answered yes. Then, obviously, one was lying and the other was telling the truth, and there would then be no way for the judge to decide which one lied, and hence which one was the spy. On the other hand, if Edwin answered no, then he agreed with Edward; hence either both lied or both told the truth, but there is no state in which they could both lie, hence they must have both told the truth and the judge could then realize that Edward was the spy. Since the judge did know, then it must have been that Edwin answered no, and the judge then convicted Edward.

Chapter XXII

A Mixed Bag: Some Logic Tricks and Games

We shall turn to some neat logical tricks and games that are both entertaining and instructive. If you've solved the rest of the puzzles in the book, this chapter should be pure pleasure. A children's game, if you will.

1 ☽ What Question?

There is a yes/no question I can ask you such that it is logically impossible for you to give the correct answer, yet the question *does* have a correct answer, and the most curious thing of all is that anyone else in the world might be able to answer it correctly! The only person in the world who definitely cannot answer it correctly is *you*!!

Can you guess what such a question might be?

2 ☽ A Mixed Bag

Once, when entertaining at a children's birthday party, I went into the kitchen and found three peaches, three plums, and three paper bags. In one bag I put two peaches, in one I put two plums, and in the remaining bag (the "mixed" bag) I put a peach and a plum. I brought the three bags into the living room, where the guests were assembled, picked three children—Andrew, Laura, and Richard—and gave each a bag. I then explained to the guests that one of the three held two peaches, one held two plums, and one was a mixed bag holding one peach and one plum, but I didn't tell them which held what. I then said to the three: "I want each of you to peek into your bag and tell the company what you have, but I want each of you to lie!" This is what they said: Andrew: "I have two peaches"; Laura: "I have two plums"; and Richard: "I have one peach and one plum."

"Very good," I said. "I see that each of you has lied, as I requested. From now on, I want each of you to tell the truth."

I then gave pencils and paper to the remaining guests and told them, "You are to figure out a strategy whereby you successively ask each of the three to pull out one fruit from his or her bag and show it to you, until you can deduce which bag is really the mixed bag. The one that comes up with a strategy involving the *minimum* possible number of necessary questions will be allowed to try it and, if successful, will win a prize.

After a time, Elizabeth, a very clever girl, came up with a minimal strategy, and it worked! What is the strategy, and what is the minimum number of necessary questions?

3 ☽

I then took John and Mary out of the room for a private huddle. We agreed that one of the two should play the role of the liar and answer all questions falsely, and the other should play the role of the truth-teller and answer all questions truthfully. Also, one of the two was to hold a quarter in his or her clenched fist, and the other was to hold up an empty clenched fist. The three of us then reentered the room, and I explained the situation, not telling which one was the liar or which one held the quarter. The aim was for the guests to determine who had the quarter. One of the guests asked John a clever question: Does the liar have the quarter? John replied, "Yes."

Was John lying or telling the truth, or is it not possible to tell? Which one has the quarter, or is it not possible to tell?

4 ☽ The Prediction Trick

The next trick I did is one anyone can do, and it is quite effective!

I wrote something on a piece of paper, then folded it and handed it to one of the boys to put in his pocket (so that I couldn't change it) and explained, "I have just written down a description of a certain event that will or will not take place in this room in the next fifteen minutes. I'll give any of you odds of two to one that you cannot correctly guess whether the event will take place. Are any of you game?"

One of the boys said, "Okay, I'll take you on." I then handed him pencil and paper and said, "If you believe that the event *will* take place, then write down 'yes,' but if you believe that the event *won't* take place, then write down 'no.' " He then wrote down one of the words, *yes* or *no*, but I didn't see what he had written. Nevertheless, I knew that regardless of which of the two words he had written, I had won the bet!

What could I have written that made me so sure?

5 ☽ A Logic Game

It now came time for me to distribute some prizes. For the first game, I placed two closed boxes on the table and explained that each box contained either a red or a black playing card and also that one of the two boxes contained a prize. The purpose, of course, was to determine which box contained the prize. Well, on the lid of each box I had written a sentence, and I explained that if the card in the box was red, then the sentence was true, but if the card was black, the sentence was false. I did not say how many of the cards were red. Here are the sentences I wrote:

Box 1	Box 2
One of the cards is red and one is black.	The prize is in the other box.

Which box contains the prize?

6 ☽ The Second Distribution

For the next trick, I placed three boxes on the table. I explained that one box contained a red card, one a black card, and the other contained a prize but no card. Sentences were written on all three lids, and I explained that a true sentence was written on the box with the red card, a false sentence on the box with the black card, and that the sentence on the box with the prize could be true or false. Here are the sentences:

Box 1	Box 2	Box 3
This box contains the prize.	The sentence on Box 1 is true.	Box 2 contains a black card.

Which box contains the prize?

7 ☽ A Metapuzzle

To conclude the afternoon, I placed three boxes before Alice, the birthday girl. One was wrapped in red, one in yellow, and one in blue. I explained that one box contained the grand birthday present, and the other two were empty. Alice's task was to determine which box contained the present. I told her that she could ask me any number of questions answerable by *yes* or *no*, and that I either would answer all her questions truthfully or would lie each time. First she pointed to the yellow box and asked me if that box contained the present. I answered her (either *yes* or *no*, but I'm not telling you which). Then, she asked if the red box contained the present, and again I answered (either *yes* or *no*). Alice, who is a very intelligent girl, then knew which box contained the present. Which box contained the present, and how did Alice know?

Some other questions arise: (1) Is it possible from what I have told you to determine whether I told the truth or lied? (2) Is it possible from what I have told you to determine what answers I gave? (3) Is it possible to determine some of the answers I *didn't* give?

SOLUTIONS

1 • WHAT QUESTION? The question I had in mind was: Will you answer *no* to this question? If you answer *yes*, then you are affirming that you answer *no*, and you are wrong! If you answer *no*, then you are denying that you answer *no*, so again you are wrong! Thus you cannot answer correctly either way!

Yet it might happen that someone else answers the question correctly. For example, you might refuse to answer, hence the *correct* answer to the question is *no*, and someone else might well give it. Or it is possible that you incorrectly answer *no*, hence the correct answer to the question is *yes*, and someone else could give it.

2 • A MIXED BAG Only one question is necessary, and it is sufficient to determine the contents of each of the three bags!

One should ask Richard (who falsely claimed to have the mixed bag) to show you one of his fruits. Suppose he pulls out a plum. Then you know that the other fruit in the bag is also a plum (since the bag is not mixed). It then follows that Andrew's bag doesn't really have two peaches, as Andrew falsely claimed, nor can it have two plums (since there are only three plums altogether, and Richard has two of them); hence Andrew must have the mixed bag (and hence also, Laura has two peaches). On the other hand, if Richard showed you a peach instead of a plum, then by symmetric reasoning, Richard would have two peaches, Laura would have the mixed bag, and Andrew would have two plums.

3 • It is not possible to tell whether John lied or told the truth, but it is possible to tell who held the quarter.

Suppose John told the truth. Then the liar really does have the quarter, and the liar is then Mary, so in this case, Mary has the quarter. On the other hand, suppose John lied. Then it is not true that the liar has the quarter, so the truth-teller has the quarter, and the truth-teller is now Mary (since John lied), so in this case again, Mary has the quarter. Thus Mary has the quarter regardless of whether John told the truth or lied. There is no way, however, of telling whether John told the truth or lied.

4 • THE PREDICTION TRICK What I wrote was: "You will write *no*."

If he writes *yes*, then he has affirmed that the event *will* take place, but it didn't (since he didn't write *no*). On the other hand, if he writes *no*, then he indicates that the event *won't* take place, but it *did* (the event being that he will write *no*). Thus, whatever he writes is wrong!

The underlying logic of this trick is really the same as that of Problem 1. (Will you answer *no* to this question?)

5 • A LOGIC GAME Suppose the card in Box 1 is red. Then it is true that one of the cards is red and one is black, hence the card in Box 2 must be black. Now, suppose the card in Box 1 is black. Then the sentence on the lid is false, and so it is not the case that one is red and one is black; they must be of the same color, hence the card in Box 2 is black (like the one in Box 1). This proves that regardless of whether the card in Box 1 is red or black, the card in Box 2 is black. Therefore, the sentence on the lid of Box 2 is false, and the prize is in Box 2.

6 • THE SECOND DISTRIBUTION Could the red card be in Box 1? No, because if it were, then the prize would be in Box 1 (since the sentence would be true). Thus, the red card is not in Box 1. Could it be in Box 3? Well, if it were, then Box 2 would contain a black card, hence its sentence would be false, hence the sentence on Box 1 would be false, hence the prize would have to be in Box 3 together with a red card, which cannot be. Thus, the red card is not in Box 3, it is in Box 2. And the prize is in Box 1 (as the sentence on Box 2 says).

Box 1 contains the prize, Box 2 has a red card, and Box 3 has a black card (since its sentence falsely says that Box 2 has a black card).

7 • A METAPUZZLE If one of my answers had been *yes* and the other *no*, then there would be no way that Alice could have known which was the right box, for suppose I had answered *yes* to the first question and *no* to the second. Then Alice would know that if I was telling the truth, then the yellow box would have the present, whereas if I were lying, the red box would be the one, but she would have no way of knowing which. On the other hand, if I had first answered *no* and then *yes*, again she

couldn't pick the box (red, if truthful, and yellow, if lying, but which?). But she *did* know, and therefore my two answers had to be the same.

Suppose I had answered *yes* both times. Then it would be obvious that I would be lying (since the prize couldn't be in *both* the yellow and red boxes), so Alice would then know to pick the blue box. Now, suppose both my answers were *no*. Then they couldn't both be lies (for then the prize would again be in both the yellow and red boxes), so I must have told the truth, hence the prize really is not in either the yellow or red box, so again it must be in the blue box.

It is not possible for you to know what answers I gave, or if I was truthful. All you can deduce is that the prize must be in the blue box and that I didn't give different answers to the two questions.

Chapter XXIII

Some Special Problems

1 ☽ A Gödelian Puzzle

Let us define a logician to be *accurate* if everything he can prove is true; he never proves anything false.

One day, an accurate logician visited the Island of Knights and Knaves, in which each inhabitant is either a knight or a knave, and knights make only true statements and knaves make only false ones. The logician met a native who made a statement from which it follows that the native must be a knight, but the logician can never prove that he is!

What statement would work?

Author's Note: This puzzle is closely related to the famous result known as *Gödel's Theorem*, in a manner that we will discuss following the solution.

2 ☽ A Follow-up

(To be read *after* the solution of the last problem.)

Suppose we are given the additional information that the logician can do logic at least as well as you and I. Now, we proved in the solution to the last problem that the native must be a knight. What is to prevent the logician from going through the same reasoning and hence coming up with the conclusion that the native is a knight? He would thus *prove* that the native is a knight, which would falsify the native's statement, hence making him a knave! Don't we thus have a paradox on our hands?

3 ☽ Another Gödelian Puzzle

Another accurate logician once visited the island and met a native who made a statement from which it follows that the native must be a knave, but it is impossible for the logician to prove that he is.

What statement would work?

4 ☽ A Doubly Gödelian Problem

Another accurate logician visited this island and came across two natives, *A* and *B*, who each made a statement from which it logically follows that at least one of them must be a knight who is not provably so by the logician, but there is absolutely no way to tell which one it is!

What two statements would work?

5 ☽ A Gödelian Machine

Here is a machine version of Gödel's construction that illustrates the essential idea in a very instructive way.

We have a computing machine that prints out various sentences constructed from the following three symbols:

P R N

By a *sentence* is meant any combination of these three symbols that is of one of the following four forms (where *x* is any combination of those symbols whatsoever):

1. P*x* (for example, PNRNP)
2. NP*x* (for example, NPRRN)
3. RP*x* (for example, RPPNRP)
4. NRP*x* (for example, NRPNNPR)

I will explain what these sentences mean in a moment.

A sentence is called *printable* if the machine can print it. The machine is so programmed that anything it *can* print, it *will* print, sooner or later.

Now, here is what the sentences mean. (As a memory guide, "P" stands for "printable," "N" stands for "not," and "R" stands for "repeat," where by the *repeat* of expression *x* is meant the expression *xx*—that is, *x* followed by itself—for example, the repeat of RPNP is RPNPRPNP.)

1. P*x* means that *x* is printable, and is accordingly called

true if and only if *x* is printable (for example, PNRP is a true sentence if and only if NRP is printable).

2. NP*x* is called *true* if and only if it is *not* the case that *x* is printable—in other words, that *x* is *not* printable.

3. RP*x* means that when *repeated*, *x* is printable—in other words, that *xx* is printable (for example, RPNNR is true if and only if NNRNNR is printable).

4. NRP*x* means the opposite of RP*x*—in other words, that *xx* is *not* printable.

This constitutes a perfectly precise definition of what it means for a sentence to be *true*. We have here an interesting loop: The machine is self-referential in that it prints out various sentences that assert what the machine can and cannot print! We are given that the machine is totally accurate, in that every sentence it prints is true; it *never* prints anything false! This has several ramifications: For any expression *x*, if the machine can print P*x*, then P*x* must be true, hence *x* will be printed sooner or later. If RP*x* is printable, then *xx* will be printed sooner or later, and if NRP*x* is printable, then *xx* will never be printed.

Now, suppose *x* is printable; does it necessarily follow that P*x* is printable? Well, *x* being printable makes P*x* *true*, but we're not given that all true sentences are printable, only that all printable sentences are true, and so we have no reason to believe that P*x* is printable on the mere basis that P*x* is true. As a matter of fact, there *is* a true sentence that is definitely *not* printable, and the problem now is to find such a sentence. (Hint: Construct a sentence that asserts that it itself is not printable [just like the knight who said that the logician could never prove that he is a knight].)

6 ☽ A Double Version

In analogy with Problem 4, one can construct *two* sentences, x and y, such that x asserts that y is printable and y asserts that x is not printable—and hence that one of them must be true but not printable, but there is no way of telling which one it is! What pair of sentences would work? (Actually there are two solutions to this.)

7 ☽ What Went Wrong

At a lecture I once gave to a group of faculty and graduate students in logic, I did the following: I showed the group two sealed envelopes and explained that one of them contained a dollar bill, and the other a folded piece of blank paper. The idea was to determine which envelope had the dollar bill. On the face of each envelope was written a sentence. Here are the sentences:

1	2
The sentences on both envelopes are false.	The dollar bill is in the other envelope.

I explained that if anyone in the audience could determine which envelope contained the dollar bill, he or she could have it! However, the person trying this must first *prove* which envelope has the bill before opening it.

At this point, can *you* determine which envelope has the bill? Well, one student in the audience said he could, and gave the following explanation: "The sentence on the first envelope (Envelope 1) couldn't possibly be true, for if it were, then both sentences would be false (as the sentence on Envelope 1 says); hence the sentence on Envelope 1 would be false, which is a contradiction. Therefore, the sentence on Envelope 1 is false. Since it is false, then it is *not* the case that both sentences are false; hence the sentence on Envelope 2 must be true, and so the dollar bill must really be in Envelope 1, as the sentence on Envelope 2 says. This proves that the bill is in the first envelope."

"That reasoning sounds pretty good, doesn't it?" I asked the audience. Most of them nodded affirmatively. "Well now, open the envelopes," I said to the student. He opened the first envelope, and the dollar bill wasn't there! He then opened the second envelope and, sure enough, there was the bill.

The audience couldn't figure out what had gone wrong with the student's argument. I certainly didn't make any false statements; all I had said was that the dollar bill was in one of the envelopes and indeed it was. So I hadn't said anything false. What, then, was wrong with the student's reasoning? (The answer involves an important principle of modern logic, and leads naturally to some of the paradoxes of the next chapter.)

What is the explanation?

SOLUTIONS

1 • A GÖDELIAN PUZZLE A statement that would work is: You cannot prove that I am a knight.

If the native were a knave, then the statement would be false, which would mean that the logician *could* prove that he is a knight, hence prove something false, contrary to the given condition that the logician is *accurate* and never proves anything false. Hence the native cannot be a knave, and so he is a knight. Since he is a knight, what he said was true, which means that the logician cannot prove that he is a knight. And so, the native is a knight, but the logician can never prove that he is!

DISCUSSION: As we have remarked, this puzzle is closely related to a celebrated result of the famous mathematician Kurt Gödel known as Gödel's Incompleteness Theorem. In 1931, Gödel took two of the most powerful known mathematical systems and showed, to everyone's surprise, that for these systems, and a whole family of other systems, there must always be sentences that, though true, are not provable in the system. Here is the essential idea behind his proof: He showed how, for each of the systems under consideration, one could assign to each sentence a number, subsequently called the *Gödel number* of the sentence, and then construct a sentence *G* that asserted that a certain number *g* belonged to a certain set *S* of numbers. Now, set *S* consisted of just those numbers that are *not* Gödel numbers of sentences provable in the system, and the number *g* was the Gödel number of the very sentence *G*! Thus, *G* asserts that its *own* Gödel number is *not* the Gödel number of a provable sentence (provable in the system, that is). In other words, *G* is true if and only if it is not provable in the system (just like the native of my puzzle is a knight if and only if the logician cannot prove that he is). Assuming the system is accurate—that only true sentences are provable—the sentence *G* must be true but not provable in the system.

2 • A FOLLOW-UP No, there is no paradox, but a very interesting truth instead! I told you that the logician could do logic as well as you and I. I also told you that the logician was accurate, but I never told you that he *knew* he was accurate! Indeed, if he could *prove* he was accurate—if, for every sentence x, he could prove that his ability to prove x implies that x must be true—if he could do that, he would become inaccurate, for he could then prove that the native is a knight (as we did), thus falsifying the native's statement, thus making him a knave.

This bears some relation to a result known as Gödel's Second Incompleteness Theorem, which is that for each of the systems of the family considered by Gödel, if the system is consistent, then it cannot prove its own consistency.

3 • ANOTHER GÖDELIAN PUZZLE A statement that works is: You can prove that I am a knave.

If the logician could prove that the native is a knave, that would make the knave's statement true, which would mean that he is a knight, and the logician would thus be inaccurate. Since we are given that the logician is accurate, then he cannot prove that the native is a knave. Thus, the native's statement is false, so he really is a knave. And so, the native is a knave, but the logician can never prove that he is.

4 • A DOUBLY GÖDELIAN PUZZLE Here is a pair of statements that work:

> *A*: You cannot prove that *B* is a knight.
> *B*: You can prove that *A* is a knight.

To begin with, *A* must really be a knight, because if he were a knave, his statement would be false, which means that the logician *could* prove that *B* is a knight, hence *B* would really be a knight (since the logician is accurate), and so *B*'s statement would be true, which means that the logician could prove that *A* is a knight, hence would be inaccurate, since *A* is a knave. Thus, since it is given that the logician is accurate, it follows that *A* must be a knight. It then further follows that, as *A* said, the logician can never prove that *B* is a knight. And so if *B* is a knight, then the logician can never prove that he is. On the other hand, if *B* is not a knight, then his statement is false, which means that the logician can never prove that *A* is a knight (even though *A* really is one).

In summary, *A* is definitely a knight. *B* could be a knight or a knave, but there is no way to tell which. However, if *B* is a knight, then the logician can never prove that he is; whereas, if *B* is a knave, then the logician can never prove that *A* is a knight.

5 • A GÖDELIAN MACHINE For any expression *x*, the sentence NRP*x* asserts that the repeat of *x* is not printable. Taking NRP for *x*, we see that NRPNRP asserts that the repeat of NRP is not printable, but the repeat of NRP is the very sentence NRPNRP! And NRPNRP is true if and only if it is not printable. Thus, NRPNRP is either true and not printable, or false and printable. The latter alternative is ruled out since the machine never prints anything false. Therefore, NRPNRP is true but the machine cannot print it.

DISCUSSION: In the various mathematical systems in use at the time of Gödel's discovery, it had been previously assumed that truth and provability coincided—that is, every sentence provable in the system was true, and all true sentences were provable in the system. Now, the machine considered in this problem is more than a mere toy, because if *S* is any of the mathematical systems subject to Gödel's argument, it is possible to translate each sentence X of the machine language to a sentence X* of the system *S* in such a way that: (1) X is true if and only if X* is a true sentence of *S*; (2) X is *printable* by the machine if and only if its translation X* is *provable* in the system *S*. Then, since NRPNRP is true but not printable, it follows that its translation NRPNRP* is true but not provable in the system of *S*! This translation NRPNRP* is Gödel's famous sentence that asserts its own nonprovability in the system and (assuming that the system is *correct* in that no false sentences are provable) is true but not provable in the system. Thus Gödel showed that each such system is *incomplete*, in that not all true sentences of the system are provable in it!

6 • A DOUBLE VERSION We want sentences *x* and *y* such that *x* asserts that *y* is printable, and *y* asserts that *x* is not printable. One solution is to take *x* = PNRPPNRP and *y* = NRPPNRP. Well, *x* asserts that NRPPNRP (which is *y*) is printable, and *y* asserts that the repeat of PNRP, which is PNRPPNRP, which is *x*, is not printable.

Another solution is to take *x* = RPNPRP and *y* = NPRP-NPRP. Then *x* asserts that the repeat of NPRP, which is *y*, is printable, and *y* asserts that RPNPRP, which is *x*, is not printable.

7 • WHAT WENT WRONG The error was the assumption that the sentence on Envelope 1 was either true or false! Not every sentence is either true or false—for example, the well-known paradoxical sentence: This sentence is false. That sen-

tence cannot be either true or false without contradiction (see the next chapter for a discussion). And so it is with the sentence on Envelope 1. If it were true, then you would have an obvious logical contradiction, whereas the only way it could be false would be if the dollar bill was in Envelope 1, which it wasn't. Therefore, that sentence couldn't be false either. (The sentence on Envelope 2 is either true or false; it happens to be false!)

This leads us naturally to the fascinating subject of paradoxes.

Chapter XXIV

Some Strange Paradoxes!

Paradoxes have intrigued the human mind since time immemorial. We will consider several of them, some old and some new. One of the oldest is the well-known one about the Cretan who said, "All Cretans are liars." Actually, this is really not a paradox, for reasons I will shortly explain. A better version is the following sentence:

This sentence is false.

Is this sentence true or false? If it is true, then what it says is the case, which means that it really is false, as the sentence says. But this is clearly a contradiction! On the other hand, if the sentence is false, then what it says is not the case, which means that the sentence is not really false, so again we have a contradiction. And so, either way, we have a contradiction, and thus we have a paradox.

The weakness of the Cretan version is this: First of all, what is meant by a liar—one who sometimes lies, or one who always lies? If the former, we certainly have no paradox, so let us assume that *liar* here means one who always lies. Do we then have a paradox? No, we don't: The statement itself can't be true (or we would have a contradiction); the statement must be false. This means that the speaker lies sometimes (since he made a false statement) and also that at least one Cretan *sometimes* tells the truth. That is the correct conclusion, and it is no paradox. It is like a person who believes that all his beliefs are wrong! The correct conclusion is not that he is inconsistent, but more amusingly that *that* belief is wrong, and hence that at least one of his beliefs is right! So if a person believes that all of his beliefs are wrong, then at least one of his beliefs is right.

A variant of "This sentence is false," which I am fond of using, is:

You have no reason to believe this sentence.

Do you have reason to believe it, or don't you? I must tell you of an amusing incident: Some time ago, I gave a lecture on puzzles and paradoxes at a university, and sometime before the lecture, I entered the hall and wrote the above sentence in large letters on the blackboard to give the entering audience something to mull over. When I finally came in, I saw a very bright-looking boy, nine years old, sitting in the front row, and I pointed to the sentence and asked him, "Do you believe this sentence?" He replied, "Yes." I then asked him, "What is your reason?" He stunned me by cleverly saying, "I don't have any." I then asked, "Then why do you believe it?" He replied, "Intuition." He really outsmarted me!

These paradoxes, frivolous as they might seem, are closely related to very serious paradoxes that, at the turn of the century, threatened to show that mathematics was inconsistent with logic! One of the great works on the foundations of mathematics was that of Gotlob Frege, which developed a great deal of set theory from the one basic axiom that given any property, there exists the set of all things having the property. This seems reasonable enough, but as shown by Bertrand Russell, this harmless-looking axiom actually leads to an inconsistency: Call a set *ordinary* if it is not a member of itself. For example, the set of all people is itself not a person, so this set is ordinary. The set of all chairs is another example of an ordinary set. A set that *is* a member of itself is called *extraordinary*. For example, one might say that the set of all sets, being itself a set, is a member of itself, and is hence extraordinary. Whether extraordinary sets really exist or not might be questioned, but ordinary sets certainly do exist. Virtually all sets we ever consider are ordinary. Then by Frege's axiom, we can talk about the set of all ordinary sets—call this set Z. Thus every member of Z is an ordinary set and every ordinary set is a member of Z. Thus a set is ordinary if and only if it belongs to Z. Now, is Z ordinary or not? If it is, then it belongs to Z (since all ordinary sets do), but Z belonging to Z would make Z extraordinary! Thus, it is contradictory to assume that Z is ordinary. On the other hand, suppose that Z is extraordinary. This means that Z belongs to Z (Z belongs to itself), but this is not possible since only *ordinary* sets belong to Z! Thus it is also impossible that

Z is extraordinary. So whether *Z* is ordinary or not we have a contradiction. This is Russell's famous paradox.

What should we make of this? Sometime around 1916, Russell gave a popularized version of this, known as the *barber paradox*: A barber in a certain town shaved all and only those inhabitants who didn't shave themselves—that is, he shaved *every* inhabitant of the town who didn't shave himself, but he never shaved any inhabitant who did shave himself. Did the barber shave himself or didn't he? If he did, then he was one of the inhabitants who shaved himself, but he would never shave such an inhabitant! If he didn't shave himself, he was one of the inhabitants who didn't shave himself, but the barber shaved all such inhabitants, so we again have a contradiction.

What is the solution to the barber paradox? The solution is so obvious that almost everyone overlooks it! To give you a hint, suppose I told you that a certain man is more than six feet tall and less than six feet tall. How would you explain that? What your explanation *should* be is that I am either mistaken or lying! There clearly can't be such a man! And likewise, there simply can't be such a barber. Yes, the solution is that simple! But with the Russell paradox, the matter is more serious, because it intuitively seems that there really is such a thing as the set of all ordinary sets, but in actuality there can't be, since it leads to a contradiction! So Frege's fundamental axiom—that for any property there exists the set of all things having that property—intuitive as it seems, must be given up. In place of it, various set theories have arisen with weaker axioms that are less elegant but presumably consistent.

The Russell paradox has many variants. A neat one is the Grelling paradox, which is this: As numbers get larger and larger, more words are usually required to describe them. And for any fixed number of words, there must always be a *least* number not describable in less than that number of words. In particular, consider the number described as follows:

The least number not describable in less than eleven words.

The above is a perfectly good description of a number, isn't it? But, horror of horrors, if you count the number of words in that description, you will see that it is ten!

The Russell paradox has also had some humorous variants. For example, Quine's machine, which works only when it is out of operation. I also like the following, compliments of Lisa Collier, which she calls *the businessman's paradox*: A president of a certain firm offered a reward of $100 to any employee who could provide a suggestion that would save the company money. One employee wrote: "Eliminate the reward!"

Here is a little paradox of mine: Consider the Island of Knights and Knaves, where knights make only true statements and knaves make only false ones, and where every inhabitant is either a knight or a knave. On this island, no inhabitant can possibly say, "I am not a knight," because no knight would falsely say that, and no knave would truthfully say that. Hence, no inhabitant can say that. Now, suppose you visit this island and meet a native who says to you, "You will never know that I am a knight." You realize that if you ever know that he is a knight, that would falsify his statement, making him in fact a knave, which is not possible. Hence, you will never know that he is a knight. Well, he said just that, so he spoke the truth, and hence he must be a knight. Now you know he's a knight! But he said that you never would, hence he's a knave. So he is both a knight and a knave, which cannot be!

The above is a purified version of the paradox of the surprise examination: On Monday morning, a professor says to the class, "I will give you an examination someday this week—Friday, at the latest—but it will be a *surprise* examination—that is, you will not know that you will get it any time before you get it." Well, a bright student reasoned thus: "I can't get it on Friday, because if at the end of Thursday's class I haven't yet gotten it, then I'll know it must be the next day, and so it won't be a surprise. This rules out Friday. Thus, I will get the exam by Thursday at the latest, but by the same reasoning, it can't be Thursday, because if by the end of Wednesday I haven't gotten the exam, I'll know it must be Thursday, since I have already proved that Thursday is the last possible day. But then the Thursday exam will be no surprise. This rules out Thursday, so Wednesday is the last possible day." The student then ruled out Wednesday (by a similar argument), then Tuesday, then Monday, and so concluded: "Therefore, I can't get the exam at all! Just then, the professor said, "Now I will give you your exam." The student was most surprised!

A very baffling paradox of recent origin is the *Newcombe* paradox, which I will first state and then give a new variant. Suppose I show you a chest with two drawers and explain that each drawer contains either a hundred-dollar bill or a thousand-dollar bill. I then give you the option of either taking whatever there is from both drawers or taking just the money in the bottom drawer. Which would you choose? Almost everyone will say: "Obviously, I should take both drawers, since I will get more than if I chose just one drawer—more by either $100 or $1000." Well, is there *any* further information I could give you that would make you change your mind and believe that you will find more money if you open just the bottom drawer than if you open both drawers? This sounds ridiculous, doesn't it? I'm sure that many of you would be willing to bet that *no* added information could make you change your mind on *that*! But wait! What I didn't tell you is that there exists a being—either a human, a computer, or a god—who is a *perfect* predictor and at any time knows the entire future of the universe. This perfect predictor knew in advance how you would choose, and was in control of how much money was to go into the drawers. If the being predicted that you would choose both drawers, then only $100 was to be put into each drawer, but if it was determined that you would choose just the bottom drawer, then $1000 was to be put in each drawer. Would this added information change your mind? I'm sure that many of you will say *yes*, on the simple grounds that you now realize that if you open both drawers, you will find only $200 (as the predictor arranged), whereas, if you open only the bottom drawer, you will find $1000. But yet, doesn't this contradict the equally obvious fact that the money is already *there*, and there is twice as much in both drawers together than in the bottom drawer alone? This is the paradox.

What is the solution? Some have proposed that this paradox proves that there cannot be such a thing as a perfect predictor. I don't agree with that solution at all! In fact, it is possible to reformulate the essential idea behind the paradox leaving the predictor out entirely! So here is my variant.

Again there is a chest with two drawers and either there is $100 in each drawer or $1000 in each drawer, and again you are to choose either both drawers or just the bottom drawer. Suppose we are also given the following:

Proposition: Either you will choose both drawers and there will be $100 in each, or you will choose just the bottom drawer and there will be $1000 in it (as well as $1000 in the top drawer).

Note that I have said nothing about any predictor! (Newcombe's version with the predictor *implies* the above proposition, but the above proposition is more general and makes no reference to a predictor.) Now, is the above proposition consistent? Well, I will first prove that the proposition is inconsistent, and then I will prove that it is consistent. This is my version of the paradox.

Proof that the proposition is inconsistent: Regardless of whether there is $100 in each drawer or $1000, the fact remains that in either case, there is more money in both drawers than there is in just the bottom drawer. Therefore, if you open both drawers, you will find more money than if you open only the bottom drawer. On the other hand, it follows from the proposition that if you open just the bottom drawer, you will find more money ($1000) than if you open both drawers ($200). This is clearly a contradiction, and so the proposition is inconsistent.

Proof that the proposition is consistent: If there is any possibility that the proposition could be true, then the proposition must be consistent (since an inconsistent proposition can never be true). Now, it certainly is possible that you open both drawers and find $100 in each one, in which case the proposition is validated. (Also, it is possible that you open just the bottom drawer and find $1000 in it—and later find out that the top drawer also has $1000—which again validates the proposition.) Since the proposition *can* be validated, then it must be consistent.

And so, I have proved that the proposition is both consistent and not consistent, and that certainly is a paradox, isn't it?

Next, I wish to give a paradoxical version of what is known as the *prisoner's dilemma.* This is not traditionally stated as a paradox, but I will show how it can be converted into one. I will consider the *positive* version, in which the participants are rewarded rather than punished: Let us say that you and I are the players and there is also a rewarder. You and I have two options: to *cooperate* with each other, or *defect.* If we both cooperate, we are each rewarded with $3. If we both defect, we each get $1. But if one of us cooperates and the other defects, the defector gets $5 and the cooperator gets nothing! What is your best strategy? Well, if I cooperate, you will get more by defecting than by cooperating ($5 versus $3). Also, if I defect, you will get more by defecting than cooperating ($1 versus nothing). So regardless of what I do, you are better off defecting. Therefore, you should defect. By the same token, I should defect. And so we both defect, hence we each get $1, whereas if we had both cooperated, we would each have got $3! And so the strange thing is that we are better off if we both cooperate, yet each of us is individually better off defecting! Another way to look at it is this: Assuming that you and I are rational creatures, then, since the conditions between us are perfectly symmetrical, we will play alike. Knowing that we will play alike, we each should obviously cooperate (thus earning $3 as opposed to $1). Yet, by the first argument, each of us should defect!

Let us assume that we will play alike. Then the following four propositions hold:

Proposition 1: If we both cooperate, then we each get $3.

Proposition 2: If we both defect, then we each get $1.

Proposition 3: If one of us cooperates and the other defects, the defector gets $5 and the cooperator gets nothing.

Proposition 4: We will play alike—that is, either we will both cooperate or both defect.

Are these four propositions consistent? I will first prove that they are not; then I will prove that they are—thus getting a paradox of the same form as the preceding one. Well, to prove them inconsistent, it follows from just the first three propositions alone that you will get more by defecting than by cooperating, for regardless of what I do, you will get more by defecting, as shown earlier. But adding Proposition 4, it follows that you will get more if you cooperate ($3 versus $1). This is clearly an inconsistency, so the four propositions cannot be consistent. On the other hand, the propositions *must* be consistent, because it is possible that we both play alike, and if we do (either by both cooperating or both defecting), all four propositions will be validated. And so the propositions are consistent after all. Yet I have proved that they are not! This is the paradox.

Lastly, I wish to consider what is known as the *envelope paradox*, plus a variant of it: On the table are two envelopes, one of which contains twice as much money as the other (and neither one is empty). You pick one of the envelopes and open it, and you then have the option of either keeping the contents, or trading it for the money in the other envelope. Are you probably better off trading? Well, suppose you find, say, $100 in your envelope. Then the other envelope contains with equal probability either $200 or $50, so if you gain by trading, you will move up from $100 to $200, thereby gaining $100, whereas if you lose by trading, you will lose only $50 (moving down from $100 to $50), and since the probabilities of gaining or losing are equal, you should trade. In other words, you are not playing for double or nothing, which is a fair bet (neither favorable nor unfavorable), but for double or *half*, which is definitely a *favorable* bet. Therefore, you should trade. But even before you open your envelope, you know that whatever amount you find, you can reason the same way and conclude that you should trade, and so without even bothering to open the envelope, you should trade right away! But that is clearly ridiculous!

What is the solution? Well, the solution usually given by experts in probability is that there is no such thing as a probability measure on the infinite set of positive whole numbers. But I maintain that probability is really quite inessential to the heart of this paradox, and to prove this, I will now present a variant of the paradox in which probability is entirely eliminated.

Again, there are two envelopes, one of which contains twice as much money as the other. You are holding one of the two envelopes in your hand and have already decided to trade it for the other. I will now prove to you the following two *logically incompatible* propositions.

Proposition 1: The amount you will gain by trading, if you do gain, is greater than the amount you will lose, if you do lose.

Proposition 2: The two amounts are really the same.

To prove Proposition 1, let n be the amount you are now holding. Then the other envelope either contains $2n$ or $n/2$. If you gain by trading, you will gain n dollars (moving from n to $2n$), whereas if you lose by trading, you will lose only $n/2$ (moving down from n to $n/2$). Since n is greater than $n/2$, then Proposition 1 is established.

To prove Proposition 2, let d be the difference between the two amounts in the envelopes (or what is the same thing, the lesser of the two amounts). Well, if you gain on the trade, you will gain d dollars. If you lose on the trade, you will lose d dollars. Since d is equal to d, then Proposition 2 is established!

Now, Propositions 1 and 2 obviously can't *both* be true! Which of the two do *you* believe?

Actually, we could leave out the envelopes altogether! Here is a very pure form of the paradox.

Suppose we have two positive whole numbers, x and y, and all we are told about them is that one of them is twice the other. Now, I shall prove the following two logically incompatible propositions.

Proposition 3: The excess of x over y, if x is greater than y, is greater than the excess of y over x, if y is greater than x.

Proposition 4: The two excesses are the same.

Proof of Proposition 3: Either $x = 2y$ or $x = \frac{1}{2}y$. If $x = 2y$, then the excess of x over y is $x - y = 2y - y = y$. On the other hand, if $x = \frac{1}{2}y$, then the excess of y over x is $y - x = y - \frac{1}{2}y = \frac{1}{2}y$. Clearly, y is greater than $\frac{1}{2}y$, which proves Proposition *3*.

Proof of Proposition 4: Let d be the difference between the two quantities x and y (or what is the same thing, the lesser of the two). If x is greater than y, then the excess of x over y is d. If y is greater than x, then the excess of y over x is again d. Since $d = d$, then the two excesses are obviously the same.

Hmm!

Solutions to the Scheherazade Puzzles

Chapter II

1 • WHAT IS IT? The answer is *Nothing*.

2 • WHAT ARE THE CHANCES? The chances are 100 percent. It is *certain* that at least two Arabs must have exactly the same number of Arab friends. Let us say there are a million Arabs. Then the possible number of Arab friends for a given Arab ranges from 0 to 999,999. If no two Arabs had the same number of Arab friends, then each of these million numbers would have to be realized—that is, one Arab would have to have 0 Arab friends; one would have to have just 1 Arab friend; one, 2 Arab friends, and so forth up to one Arab having 999,999 Arab friends. But if an Arab has 999,999 Arab friends, he has *all* the other Arabs as friends, hence every Arab would have to have him as a friend, so no Arab could have 0

Arab friends. Thus, the assumption that every Arab has a different number of Arab friends leads to a contradiction; hence, at least two Arabs must have the same number of Arab friends. (Of course, the number *one million* was chosen arbitrarily; the same kind of proof would work with any other number, assuming, of course, that there are at least two Arabs.)

3 • HOW DID THEY MANAGE? The two camels were facing each other the whole time, hence facing in opposite directions.

4 • ABDUL THE JEWELER The usual answer is 6, but the job could be done cutting only 5 links, because each chain has only 5 links. The jeweler could take one chain completely apart and, with those 5 loose links, attach the remaining 5 chains into a circle.

5 • THE SECOND TALE OF ABDUL THE JEWELER It's easiest to work the problem backwards: The fourth thief must have found 2 diamonds, the third thief 6, the second thief 14, and the first thief 30.

6 • TWO OTHER VERSIONS If the second version is correct, Abdul had 60 diamonds. If the third version is correct, then Abdul had 76 diamonds.

7 • ABDUL AND THE TEN THIEVES This can be solved by trial and error, or by algebra, but I like best the following commonsense solution:

First, let each one take 5 pearls. There are then 6 pearls left over, and the junior members have already received their portion, so the 6 remaining pearls must go to the senior men, so there are 6 senior thieves.

8 • HOW MANY? We can solve this by subtracting 9 from 59, then 9 from the result, then 9 from that result, and so forth until we hit a multiple of 4. So we begin: 59, 50, 41, 32. Well, 32 is divisible by 4. And so there were 32 rubies in 8 bags, and 27 emeralds in 3 bags.

9 • A SIMPLE ONE? No, they are not equal. Six dozen dozen is 6×144, which is 864; whereas, a half dozen dozen is 6×12, which is 72.

10 • SINBAD AND HINBAD The answer is not 6, but 3.

11 • SINBAD The usual answer, the fourth rung from the bottom, is wrong. The correct answer is the second rung from the bottom, because the boat rises with the water.

12 • HOW MANY PONIES? He had 120 ponies. This could be found by trial and error, or by solving the equation $x/4 + x/3 = 10 + x/2$.

13 • THE PONY WHO GOT LOST Fifty-five miles in 5 days averages out to 11 miles per day. Now, since the pony's increase was regular (one mile each day), he must have hit his exact average on the third day. So he walked 11 miles on the third day. This means that he walked 9 miles the first day, 10 miles the second, 11 miles the third, 12 miles the fourth, and 13 miles the fifth, or last, day. If you check, you will find these five numbers add up to 55.

14 • THE MAGIC TREE No, the answer is 99 days. Thus, on the ninety-ninth day, the tree was half its full height, and on the next day, it doubled and became its full height.

15 • ANOTHER MAGIC TREE Let's say that the tree was originally one foot tall. On each day, it grew half a foot (since on the first day it grew one half a foot; on the next day, one third of 1½ feet, which is half a foot; and so forth). So, in 198 days it grew 99 feet, and then was 100 feet tall. Thus the answer is 198 days.

CHAPTER III

16 • HASSAN'S HORSES One would think that the answer would be 7—the 4 white ones and the 3 black ones—but the answer is *none*, since horses can't talk.

17 • HOW MUCH? A million divided by one fourth is 4 million, not a quarter of a million! A million *times* one fourth is a quarter of a million, but a million *divided* by a fourth is the number of fourths that go into a million. Think of it this way: How many quarters do we need to make a million dollars? Or, remember the old rule that to divide by a fraction, you invert and multiply.

So, the answer to the problem is 4,000,050.

18 • HASSAN'S HORSES AGAIN Scheherazade was right. If those horses could talk, then all of them *could* say it, but the brown horse would have to be wrong.

19 • HASSAN'S MULE Letting x be the mule's age, we have $x + 4 = 3(x - 4)$, which makes $x = 8$.

20 • WHAT COLOR? If the mule were black, all three guesses would be wrong. If the mule were brown, all three guesses would be correct. Therefore, the mule must be gray (and thus the first two guesses were correct and the third guess was wrong).

21 • HASSAN'S CAMELS Since all *but* 5 died, then those 5 were left, and so indeed any *dunce* might well give the wrong answer *3*.

22 • HOW MANY WIVES? The younger brother had 2, the older brother had 3, and the uncle had 4.

23 • HOW TALL IS THE PLANT? Solving the equation $x + 3 = 2(x - \frac{1}{2})$, we get $x = 4$.

24 • HOW TALL ARE THE FLOWERS? The red one is 10 inches tall and the blue one is 3 inches, as can be found either by trial and error, or by solving the simultaneous equations $r = 7 + b$, and $r - 4 = 2b$ (r for red; b for blue).

25 • HOW FAR? Since the cat trotted back twice as fast as she went, she spent twice as long going as returning. This means she spent 10 minutes going and 5 minutes returning. So, she went away from home at the rate of 3 miles an hour, which is one mile in 20 minutes. But since she walked only 10 minutes, she walked half a mile.

Of course, this could also be solved algebraically: Letting d be the distance, the time she spent walking is $d/3$, and the time she spent trotting back is $d/6$, and since she was gone a quarter of an hour, d is determined by the equation $d/3 + d/6 = 1/4$

26 • HOW MANY MICE? There were 27 mice. The problem is most easily solved by going backwards: 8 is two thirds of 12, which is two thirds of 18, which is two thirds of 27.

27 • ALI AND HIS PETS Ali had 7 cats and 5 dogs. After the first magician transformed a cat into a dog, Ali then had 6 cats and 6 dogs. The next day, he was back to 7 cats and 5 dogs. Then the next day, he had 4 dogs and 8 cats, which is twice as many cats as dogs.

This answer could be found by trial and error, or by solving the simultaneous equations $c - 1 = d + 1$, and $c + 1 = 2(d - 1)$, where c is the number of cats and d is the number of dogs.

28 • AUTHOR'S NOTE: He got so little money because the cat was only an Ali-cat.

29 • THE WISDOM OF HAROUN EL-RASHID This is most easily solved by dividing each loaf into 3 equal pieces. Thus, there were 24 pieces altogether; Ahmed had 15 pieces and Ali had 9. These 24 pieces were evenly divided among the three people, so each one had 8 pieces. Thus Ahmed contributed 7 pieces and Ali contributed only one piece, so Haroun was right.

30 • A SEQUEL Again, by dividing each loaf into 3 equal parts, we see that Ali had 9 pieces, Ahmed had 6, and since each of the three ate 5 pieces, Ali contributed 4 pieces and

Ahmed contributed 1. Thus, Ali should get 8 coins and Ahmed 2.

CHAPTER IV

31 • FIRST NIGHT

4	1	4
1		1
4	1	4

32 • SECOND NIGHT

2	5	2
5		5
2	5	2

33 • THIRD NIGHT

1	7	1
7		7
1	7	1

34 • FOURTH NIGHT

	9	
9		9
	9	

35 • FIFTH NIGHT

5		4
4		5

36 • AN ANCIENT PUZZLE The answer could be obtained by solving the equation $x/4 + x/5 + x/3 = x$, but I prefer the following commonsense solution: One fourth, plus one fifth, plus one third, is forty-seven sixtieths, so the remaining thirteen sixtieths of his life is 13 years. Well, thirteen sixtieths of what is 13? Obviously, 60.

37 • ANOTHER ANCIENT ONE Let x be the number of coins and y the number of beggars. Then we are given the two equations $7y = x - 24$ and $9y = x + 32$. Subtracting both sides of the first from both sides of the second, we have $2y = 56$, hence $y = 28$. Then $x = 220$.

38 • THE PUZZLE OF AHMES Adding the number to its seventh part gives us 8/7 of the number, so 8/7 of the number equals 19. Hence the number is 7/8 of 19, or 16 5/8.

Of course, the number can also be found by solving the algebraic equation $x + x/7 = 19$.

39 • A HINDU PUZZLE The number is 28. This is most easily found by reversing the whole process—starting with 2, multiplying by 10, subtracting 8, squaring the result, and so on.

40 • A SWARM OF BEES There were 72 bees. The only way I know to solve this problem, other than trial and error, is to use a quadratic equation. If we let x be the number of bees, we have the equation $\sqrt{x/2} + 2 = 1/9x$ (since one ninth of the swarm is out). Letting $y = \sqrt{x/2}$, we have $x = 2y^2$, and the equation becomes $y + 2 = 2/9y^2$. The only positive solution is $y = 6$, hence $y^2 = 36$ and $x = 72$.

41 • MORE BEES This is easier, since it involves only the linear equation $(x/5 + x/3) + 3(x/3 - x/5) + 1 = x$, whose solution is $x = 15$.

42 • TWO REPORTS Let us determine how many bees there are altogether according to the first report, and then according to the second report.

For the first report, we first note that every bee is of exactly one of the following 8 types:

1. large yellow male
2. large yellow female
3. large brown male
4. large brown female
5. small yellow male
6. small yellow female
7. small brown male
8. small brown female

We will have to find out how many bees there are of each of these 8 types. Then, since no bee is of more than one of these types, we add the 8 numbers together and this gives us our answer.

1. We are given that only one bee is of this type.
2. Since there are 4 large yellow bees and just one of them is male, 3 of them are female, of Type 2—large yellow female to bees.
3. Since there are 3 large males and just one of them is yellow, there are then 2 large brown males.

4. We have counted 6 large bees so far (1 yellow male, 3 yellow females, 2 brown males), hence the remaining 7 of the 13 large bees must be large brown females.
5. Since there are 5 yellow males and only one of them is large, there must be 4 small yellow males.
6. Of the 14 yellow bees, one is of Type 1, 3 are of Type 2, and 4 are of Type 5—which totals 8. The rest of the yellow bees must be of Type 6. And so there are 6 bees of Type 6.
7. Of the 12 males, one is of Type 1, 2 are of Type 3, and 4 are of Type 5. This totals 7. The remaining 5 male bees must be of Type 7.
8. There are no bees of this type, since we are given that every bee is either large, male, or yellow.

When we total the number of bees of each type, we find 28 bees altogether. So, if the first report is correct, there are 28 bees.

As to the second report, this is easier to figure out: Letting x be the number of bees, we have the equation $x/2 + x/4 + x/7 + 3 = x$, whose solution is $x = 28$. Thus the two reports are perfectly consistent, and so there is no reason to doubt either one.

Chapter V

43 • THE THREE CHESTS The king was wrong and made a common error. The correct answer is that the probability is 2 out of 3, though I will have trouble convincing some of you.

Let's look at it this way: The chest with the two emeralds is really out of the picture, so we can forget about it completely. We thus consider the *R R* chest (both rubies) and the *R E* chest (ruby, emerald). Let R_1 be the ruby in the top drawer of the *R R* chest, and R_2, the ruby in the bottom drawer of the *R R* chest. Let R_3 be the ruby in the *R E* chest. Now, if you pick one of the four drawers at random and find a ruby, then the chances that it is R_1, R_2, or R_3 are equal, so the chances are 2 to 1 that it is *either* R_1 or R_2. Thus, the probability is 2 out of 3 that the other drawer in the same chest is a ruby.

Another way to look at it is this: The emerald is with equal probability in any of the four drawers. If you pick a drawer and find a ruby, then the emerald is with equal likelihood in any of the other three drawers, and so *whichever* second drawer you open—whether in the same chest or in the other one—the chances of finding an emerald are only 1 out of 3, and so the chances for a ruby are 2 out of 3.

If you are still unconvinced, I suggest you try the following experiment: Take 3 red cards from a deck (to represent the rubies) and 1 black card (to represent the emerald); shuffle the 4 cards so you don't know where the black one is. Then, put the 4 cards face down on the table in two piles. Suppose you now pick one of the 4 cards, turn it over, and find it to be red. You really believe that the chance of the other card in that pile being red is one half?

If you are *still* not convinced, I suggest that you try the experiment 60 times, and see whether the number of times the other card is red isn't close to 40!

44 • THE TEN CHESTS There are ten ways in which the king can find a diamond, and with each, there are two ways of picking a second drawer of the same chest, and so there are twenty ways of first picking a diamond and then picking a

second drawer of the same chest. In how many of these ways will he pick a second diamond? Well, if he is in Chest 4, 5, or 6, there are no ways. Now, there are two ways if he were in Chest 3 (since there are two diamonds in this chest), and with each, there is only one way he could pick a second diamond, and so there are two possibilities for Chest 3. Similarly, there are two possibilities for Chest 2. As for Chest 1, there are three diamonds that he could have initially picked, and with each, there are two possibilities for a second diamond, which makes 6 possibilities altogether for Chest 1. And so there are 10 possibilities altogether for picking a second diamond, once he has picked a first. Therefore, if he has initially picked a diamond, the probability of picking a second one from the same chest is 1 in 2.

45 • TWO VARIANTS After picking a diamond initially, let us see how many ways there are of picking a diamond from a drawer of a different chest.

If he initially chose from Chest 1, he would have 3 ways, and with each one, there would be 7 remaining diamonds in the other chests, which means that there are 21 possibilities with Chest 1 as the initial choice.

For Chest 2, there are two choices, and with each, there are 8 ways he could pick another diamond from a different chest, which makes 16 possibilities. Likewise, with Chest 3, there are 16 ways. With Chest 4, there is only 1 way he could have chosen, and then there would be 9 ways he could pick another diamond from a different chest. The same is true for Chests 5 and 6. In total, there are 80 ways he can pick a second diamond from a different chest.

Now, in the first variant, there are 9 other chests, hence 27 drawers from which to pick, and since there are 10 ways he could have picked the first diamond, there are 270 ways he could pick a drawer from a different chest, after having found a first diamond. Of these 270 ways, 80 of them yield a second diamond (as we have seen), so the probability of finding a second diamond from a different chest (after having found the first diamond) is 80 out of 270, or 8 out of 27, which is considerably less than 1 out of 2. And so, he should stick to the same chest.

As for the second variant, after the last 4 chests are removed, there are 6 chests left, hence 15 drawers from a different chest than the one he first picked, and since there are 10 ways he could have found a diamond in the first place, then there are 150 ways he could initially find a diamond and then choose a drawer from a different chest. Of these 150 ways, there are 80 in which he finds a second diamond (as we already know), and so the probability is then 8 out of 15, which is slightly better than 1 out of 2. So if the last 4 chests are removed, he is better off picking his second drawer from a different chest.

46 • WHAT ARE THE ODDS? The usual wrong answer is 50–50, whereas the correct answer is 1 out of 3. It is perhaps easiest to see this with tosses of a coin.

Suppose you toss a coin twice. What are the four possibilities? They are *H H* (heads, heads), *H T* (heads, tails), *T H* (tails, heads), and *T T* (tails, tails). These four possibilities are equally likely. Now, suppose we are told that at least one is a head. This rules out *T T*, and so we are left with the three equally likely possibilities *H H*, *H T*, and *T H*, in only one of which do we have both heads! And so, if at least one is a head, the probability that both are heads is 1 out of 3, not 1 out of 2.

If you have any doubts, try making 2 tosses about 100 times, and among those in which at least one is a head, see how many times both are heads. It should be about 1 out of 3.

47 • WHAT ARE THE ODDS? No, the king was wrong! This time the odds are 50–50. Again, it might be easier to see it in terms of tossing a coin.

Again, we make two tosses. This time, instead of being told that at least one is a head, we are told that the first one is a head. Then, the chances that both are heads is the same as the chances that the second one is a head, which of course is 1 out of 2. Or, if we are told that the second one is a head, then again the chances that both are heads is 1 out of 2. But each of these is very different from being told that *at least one* is a head.

And so it is with cats: Being told that at least one is male means that there are 3 equally likely possibilities:

1. White one is male and black one is male.
2. White one is male and black one is female.
3. Black one is male and white one is female.

In only 1 of these 3 possibilities are both cats male, and so the probability is then 1 in 3 that both are male.

On the other hand, if we are told that the white one is male, then we have 2 equally likely possibilities:

1. White one male, black one male.
2. White one male, black one female.

Then in Case 1, both are male, and in Case 2, they are not both male, and so the probability now that they are both male is 50–50.

48 • A SURPRISING FACT No finite amount is enough for Ali to pay Ahmed; the expected value of the game is infinite! Ali has a 50–50 chance of winning 2 pieces of silver, and this is worth 1 piece. He also has a 1 in 4 chance of winning 4 pieces, and this is worth 1 piece. He has a 1 in 8 chance of winning 8 pieces, and this is worth another piece, and so on.

To put the matter another way: Suppose we modify the game by ruling that there are to be only 100 tosses; if no heads come up by then, Ahmed pays nothing. The expected value of the game is then 100 pieces of silver. If the maximum number of tosses is agreed to be 1 million, then the expected value of the game is 1 million pieces of silver, and so forth. If there is no ceiling on the number of tosses, then no finite amount of advance payment is adequate.

Why this is called a "paradox" is unclear. It is really no paradox; it is merely quite surprising—at least to some.

49 • A CONTROVERSIAL PUZZLE Scheherazade was right, though many people find it difficult to understand why. Even professional mathematicians have been fooled by this problem.

To begin with, the chances are 1 in 3 that the king initially chose the prize box (the box containing the prize). Now, regardless of whether the king has the prize box or not, Scheherazade, who knows where the prize is, can *always* open an empty box, and doing this really gives no additional information; the odds are still 1 in 3 that the originally chosen box contains the prize, and so the odds are 2 in 3 that the prize is in Box C. Therefore, the king should trade!

For those of you who are not convinced, just ask yourself the following: Suppose you are in the king's place and play the game a great number of times and *never* trade. How often do you expect to win? Obviously, about a third of the time—the times when you have initially chosen the prize box. On the other hand, suppose you *always* trade. Then you will win the prize about two thirds of the time—the times when you have *not* initially chosen the prize box, which is about two thirds of the time.

If you are still not convinced, suppose that the game is played with 100 boxes instead of 3, and only one of them has a prize; the rest are empty. You pick a box—say, Box 1. At this point, your chances are obviously 1 out of 100. Now, suppose someone, who knows where the prize is, deliberately opens 98 boxes and shows you that they are empty. Do you really believe that the chances that Box 1 contains the prize are now 50–50? I'll be glad to play this game with you, say 100 times, and each time I'll give you odds of 10 to 1 that the box you originally chose does *not* contain the prize—even after I've shown you 98 empty boxes!

Chapter VI

50 • ABDUL IS ROBBED AGAIN The mere fact that Shamhir lied doesn't prove his guilt, but the additional fact that the other two told the truth does, because since Sabit truthfully accused one of the others, he must be innocent, and the same with Salim. Therefore, Shamhir is indeed the guilty one.

51 • ANOTHER ROBBERY Since Ibn and Hasib contradict each other, one of them lied and the other told the truth. But

since two lied, then Abu must be one of them, hence Abu is guilty.

52 • ANOTHER ROBBERY If Hasib committed the robbery, then his statement (which agrees with Ibn's) is false, contrary to the given fact that the thief spoke the truth. Therefore, Hasib is not guilty. If Abu is guilty, then all three spoke the truth, contrary to the given condition that at least one lied. And so it was Ibn who was guilty. He spoke the truth and Hasib lied.

53 • ANOTHER ROBBERY If Hasib were guilty, then the guilty one would have told the truth, contrary to what is given. Hence, Hasib is innocent (and also lied). If Ibn were guilty, then Abu would be innocent; hence, Ibn would have told the truth, which is again contrary to what is given. Therefore, it is Abu who is guilty (and all three lied).

54 • HOW MANY? Let x be the number of emeralds. The first robber took $\frac{1}{3}x$ emeralds, which left $\frac{2}{3}x$, of which two thirds was taken by the second robber, and thus the second robber took $\frac{4}{9}x$, so the total quantity taken was $\frac{1}{3}x + \frac{4}{9}x$, which is $(3/9 + 4/9)x$, which is $\frac{7}{9}x$. Thus, $\frac{2}{9}x$ was left, so $\frac{2}{9}x = 12$. Hence, $x = 54$. So the first robber took one third of 54, which is 18. This left 36 emeralds, of which the second robber took two thirds, which is 24, thus leaving 12.

55 • A HYPOTHETICAL THEFT If one took 4 of each kind, one would have 16, and the seventeenth one would have to match one of the four, so the answer is 17.

56 • HOW MANY BAGS? He took 4 bags of 17 jewels each and 2 bags of 16 jewels each.

57 • ANOTHER ROBBERY If Ibn is guilty, then Hasib is innocent (since the sword was stolen by only one person), hence Hasib spoke the truth, which means that Abu is guilty, which again means that more than one person is guilty. Therefore, Ibn can't be guilty; he is innocent. Since he is innocent, he spoke the truth, which means that Hasib is guilty (and falsely accused Abu).

58 • ANOTHER ROBBERY Again Hasib is guilty. We leave the proof to the reader.

59 • HOW MANY? Solving the equation $3x - 10 = 2(x + 10)$, we get $x = 30$, which is the number of coins originally taken by Ibn. Hence Abu originally had 90 coins, which is three times as many. He then gave 10 coins to Ibn, and Ibn then had 40 coins and Abu had 80, which is twice as many. He must now give 20 coins to Ibn to make their shares equal.

60 • A BIT MORE GREED Abu originally had six times as many as Hasib, and so we have the equation $6x - 10 = x + 10$, which makes $x = 4$. Thus, Hasib took 4 coins, Ibn 8, and Abu 24, totaling 36.

61 • HONOR AMONG THIEVES? We work the problem backwards. Letting x be the number of coins found by Hasib, we have $\frac{4}{11}x = 8$, which makes $x = 22$. Letting y be the number of coins found by Ibn, we have $\frac{11}{16}y = 22$, making $y = 32$. Thus, Ibn took 10 coins and Hasib 14, leaving 8 for Abu.

62 • WHO STOLE WHAT? If Ibn stole the camel, then his statement that he stole neither the horse nor the mule would be true, but we are given that the one who stole the camel lied. Therefore, Ibn didn't steal the camel.

If Ibn stole the horse, then he would have lied in claiming that he stole neither the horse nor the mule, contrary to the given fact that the one who stole the horse told the truth. Therefore, Ibn didn't steal the horse. Hence, Ibn stole the mule. Hasib, then, spoke the truth in claiming that Ibn stole the mule, so Hasib didn't steal the camel (since the one who stole the camel was lying); therefore, Hasib stole the horse. If Hasib stole the horse and Ibn stole the mule, then Abu stole the camel.

63 • WHO STOLE WHAT FROM WHOM? *Step 1:* According to Fact 3, Kisra is less dangerous than the one who stole the emerald, and by Fact 1 is also less dangerous than the one who stole the diamond (who is the most dangerous); ergo, Kisra stole the ruby. Since Kisra, who stole the ruby, is the least dangerous, the one who took the emerald is the next most dangerous, and the one who took the diamond is the most dangerous.

Step 2: According to Fact 2, Amina doesn't own the emerald. Also, Kisra, who stole the ruby, didn't steal it from Amina but from the eldest of the ladies (who can't be Amina, who, per Fact 2, is younger than at least one of the other ladies). Hence, Amina doesn't own the ruby; she owns the diamond.

Step 3: Since Amina owns the diamond, then the man who stole from her is the man who stole the diamond and, hence, per Facts 1 and 4, is both a bachelor and an only child, so he could not have a brother-in-law, and therefore is not Abu. So Abu didn't steal the diamond, and since he didn't steal the ruby (Kisra did), he stole the emerald. Now we know that Kisra stole the ruby and Abu stole the emerald; so Badri stole the diamond.

Step 4: Finally, Badri stole from Amina. Since Abu didn't steal from Amina, and he didn't steal from Fatin, he stole from Safie. Thus, Abu stole the emerald from Safie, Badri stole the diamond from Amina, and Kisra stole the ruby from Fatin. This settles everything.

CHAPTER VII

64 • WHAT ARE THE AGES? A common wrong answer is 10 and 1. This is wrong because the difference in ages would be only 9, not 10. The correct answer is 10½ and 6 months or half a year old.

65 • HOW MUCH? The principle here is similar: The correct answer is not 80 pounds but 90 pounds.

66 • PEOPLE ARE NOT ALWAYS SO NICE! There were 8 men.

67 • ALI BABA'S THIEVES AGAIN It's best to work the problem backwards. We let x be the number of coins that the third thief found. Then, he took $\frac{1}{5}x = 3/5$, so what remained was $x - (\frac{1}{5}x + 3/5)$, and so $x - (\frac{1}{5}x + 3/5) = 409$, which makes $x = 512$. Next, letting y be the amount the second thief found, we have $y - (\frac{1}{4}y + 1/4) = 512$, which makes $y = 683$. Next, letting z be the amount the first thief found, we have

$z - (\frac{1}{3}z + 1/3) = 683$, which makes $z = 1025$—the amount Ibn found.

68 • A SIMPLE LOGIC PUZZLE According to Fact 1, Hassan can't be the oldest, therefore, by Fact 2, Ali is the youngest. Then, per Fact 1, since Ali isn't the oldest, Ahmed is the oldest. Thus, Ahmed is the oldest; next comes Hassan; and Ali is the youngest.

69 • WHICH ONE IS OLDER? Since the two agree, they are either both lying or both telling the truth. Since at least one is lying, then they are both lying, hence the sister is the older.

70 • A TRIAL If the first witness told the truth, then the statement of the third witness would also have to be true, contrary to the given condition that only one of the statements is true. Thus the second witness told the truth. Therefore, the third witness lied; the defendant has never committed any robberies. In particular, he did not rob the caravan, and so he is innocent. (The second witness is the only one who told the truth.)

71 • HOW FAR IS THE SHRINE? Letting x be the distance, Ali walked for $x/5$ hours and Ahmed walked for $x/4$ hours, and since Ahmed arrived a quarter of an hour after Ali, we have the equation $x/4 - x/5 = 1/4$, which makes $x = 5$. So Ali walked 25 miles and Ahmed 20.

72 • A HERMIT'S CLIMB The hermit was gone for 28 hours, and since he meditated and rested for 12 hours, he spent 16 hours walking. Letting x be the distance, the time spent walking up was $x/1\frac{1}{2}$, and the time spent walking down was

$x/4\frac{1}{2}$. And so we have the equation $x/1\frac{1}{2} + x/4\frac{1}{2} = 16$, whose solution is $x = 18$.

73 • A CLEVER STUDENT The student first paired 1 with 1000, whose sum is 1001. He then paired 2 with 999, whose sum is also 1001. Next, pairing 3 with 998, the sum is also 1001. There are 500 such pairs, the last of which is 500 and 501, whose sum is 1001. Thus, the numbers from 1 to 1000 can be paired off into 500 pairs, each of whose sum is 1001, and so the answer is 500 × 1001, which is 500,500.

More generally, for any positive integer n, the sum of the numbers from 1 to n (the sum $1 + 2 + \ldots + [n - 1] + n$) is $\frac{n}{2}(n + 1)$, because we can pair 1 with n, 2 with $n - 1$, and so forth, obtaining $n/2$ pairs each of which adds up to $n + 1$. (If n is even, the last pair is $n/2$ paired with $\frac{n}{2} + 1$, whereas if n is odd, the last pair is $\frac{n+1}{2}$ paired with itself. For example, if $n = 1001$, the last pair is 501 paired with 501. And so, whether n is even or odd, the sum is $n/2$ times $(n + 1)$, or $\frac{n \cdot (n+1)}{2}$.

The proof is more usually given as follows:

Let S be the sum $1 + 2 + \ldots + n$. Then

$$\begin{array}{rl} & S = 1 + 2 + \ldots + (n - 1) + n \\ + & S = n + (n - 1) + \ldots + 2 + 1 \\ \hline & 2S = \underbrace{(n + 1) + (n + 1) + \ldots (n + 1) + (n + 1)}_{n} \end{array}$$

Thus, $2S = n(n + 1)$, so $S = \frac{n(n+1)}{2}$.

(Incidentally, this story has been told about the mathematician Carl Friedrich Gauss.)

An entirely different proof (apparently not known these days) is attributed to Scheherazade. It is quite clever and interesting, and we will look at it after the next problem.

74 • HOW MANY WAYS? There are 1000 possibilities for Ali's number, and with each of these, there are 1000 possibilities for Ahmed's number; hence there are 1 million possibilities altogether for the two numbers. We must, therefore, compute how many ways there are in which Ahmed's number is greater than Ali's. Dividing this number by 1 million gives us the probability that Ahmed's number is greater than Ali's.

How many ways are there in which Ahmed's number is greater than Ali's? There are two different ways of computing this, and the fact that they agree has an interesting theoretical consequence, as we shall see. In the first way, we reason as follows: If Ali's number is 1, then there are 999 possibilities for Ahmed's number. If Ali's number is 2, then there are 998 possibilities for Ahmed's number, and so forth, down to Ali's number being 999, in which case there is only 1 possibility for Ahmed's number. So the total number of possibilities is $1 + 2 + \ldots + 999$, which by the formula of the last problem is $\frac{999 \times 1000}{2}$.

Here is a simpler way of computing this, and it does not use the formula of the last problem: There are 1000 ways in which the two numbers can be the same, hence there are 1 million minus 1000 ways in which they can be different, and this number is 999,000. In half these cases, Ahmed's number will be the greater, and so we again have the answer $\frac{999{,}000}{2}$.

75 • SCHEHERAZADE'S OBSERVATION We saw in the first method of solving the last problem that the number of

ways Ahmed's number can be greater than Ali's is $1 + 2 + \ldots + 999$. But by the second method, we saw that the number is $\frac{999 \times 1000}{2}$. This provides another way of seeing that the two numbers are the same!

Of course, this method generalizes to any positive n: Given numbers x and y each from 1 to n, how many ways are there in which y is greater than x? Using the first method, we see that if $x = 1$, there are $n - 1$ possibilities for y; if $x = 2$, there are $n - 2$ possibilities for y, . . .; if $x = n - 2$, there are 2 possibilities for y; and if $x = n - 1$, there is just one possibility for y, and so the total number of ways is $1 + 2 + \ldots + n - 1$. Using the second method, there are n^2 possibilities for the two numbers x and y, and in n of them, the numbers will be the same, hence in $n^2 - n$ cases the numbers will be different, and in half those cases, y will be greater than x. Thus the number of ways in which y is greater than x is $\frac{n^2 - n}{2}$, which is $\frac{(n-1) \cdot n}{2}$. This proves that for any number n, $1 + 2 + \ldots + n - 1 = \frac{(n-1) \cdot n}{2}$, or what is the same thing, for any number n, $1 + 2 + \ldots + n = \frac{n \cdot (n+1)}{2}$.

This was Scheherazade's alternative proof of this famous formula.

Chapter VIII

76 • THE MAZDAYSIANS AND AHARMANITES We are given that the two statements are either both true or both false, but they obviously can't both be true, hence they are both false. Therefore, Perviz is married and Bahman is not.

77 • ANOTHER VERSION If both statements are true, then both are unmarried. If both statements are false, then

Perviz is married and Bahman is unmarried. In either case, Bahman is unmarried. It is not possible to tell whether or not Perviz is married.

78 • A THIRD VERSION Bahman said that at least one of them is married. Suppose Perviz said that he is married. Then it could be that both statements are true, and it could be that both statements are false, and the sage would have no way of knowing which. But we are given that the wise sage *did* know, and hence Perviz must have said that he is not married. It is then not possible that both statements are false; they must both be true, and hence Perviz is not married and Bahman is married.

79 • OMAR THE MAGISTRATE This is another metapuzzle: If the defendant's second answer had been *yes*, then it would be obvious that he is an Aharmanite, but there would be no way of telling whether or not he had stolen the camel. But Omar did know, hence he must have answered *no*. Now, suppose he is a Mazdaysian. Then, as he said, he really did once claim that he never stole the camel, and being a Mazdaysian, he is innocent of the crime. But what if he is an Aharmanite? Then both his answers were lies, which means that he never claimed that he never stole the camel; he did once claim that he stole the camel, but being an Aharmanite, his claim was false, which means that he didn't steal the camel. And so, regardless of whether he was a Mazdaysian or an Aharmanite, he is innocent of the crime of having stolen the camel. (Whether he was a Mazdaysian or an Aharmanite cannot be determined.)

80 • THE TOWN CRIER *C* is obviously an Aharmanite, since no Mazdaysian would make the false statement that all three are Aharmanites. So *C* is an Aharmanite, and since his statement is therefore false, at least either *A* or *B* must be a Mazdaysian.

B is certainly not the town crier, for a Mazdaysian town crier wouldn't lie and say that the town crier is an Aharmanite, and an Aharmanite town crier wouldn't truthfully say that the town crier is an Aharmanite. And so *B* is not the town crier.

If *A* is a Mazdaysian, his statement is true, hence he is then not the town crier, and since *B* isn't, it must be *C*, who is an Aharmanite. And so, if *A* is a Mazdaysian, the town crier is an Aharmanite. On the other hand, if *A* is an Aharmanite, his statement is false, hence he is the town crier, so again, the town crier is an Aharmanite.

In summary, the town crier is either *A* or *C*, and he is an Aharmanite.

81 • BUT WHICH ONE? As we have seen, *C* is an Aharmanite. Therefore, *A*'s second statement was true, hence *A* is a Mazdaysian, and therefore not the town crier, so the town crier is *C*.

82 • WHICH ONES? Since *C* agrees with *B*, they are both of the same type (both Mazdaysians or both Aharmanites). If *B* were a Mazdaysian, then *C* would also be, but this would falsify *B*'s statement, and we would have the impossibility of a Mazdaysian making a false statement. Therefore, *B* is an Aharmanite, and hence so is *C*, and hence *A*'s statement is false. Therefore all three are Aharmanites.

83 • WHICH ONES? At most one of the 10 statements can be true, hence at least 9 of the people are Aharmanites. Now, A_{10} can't be Mazdaysian, or we would have a contradiction. And so, A_{10}'s statement is false, which means that at least one of the 10 is a Mazdaysian. Therefore, exactly 9 of the people are Aharmanites, and A_9 spoke the truth. Thus A_9 is a Mazdaysian and all the others are Aharmanites.

84 • INNOCENT OR GUILTY? A statement that would work is: The elephant was stolen by an Aharmanite. If he is Mazdaysian, he is innocent, since the elephant was then stolen by an Aharmanite, and if he is an Aharmanite, he is again innocent, since the elephant was then stolen by a Mazdaysian. Thus, he is innocent, but could be either a Mazdaysian or an Aharmanite.

85 • ANOTHER TRIAL A statement that would work is: I am not a Mazdaysian who has stolen the elephant.

If the speaker were an Aharmanite, then it would obviously be true that he is not a Mazdaysian who has stolen the elephant (in fact, he is not a Mazdaysian at all). Hence, an Aharmanite would have made a true statement, which cannot be. Therefore he must be a Mazdaysian. It then follows that his statement is true, and so he never stole the elephant.

86 • THE NEXT TRIAL A statement that works is: I am an Aharmanite who has stolen the elephant.

I leave the proof to the reader.

87 • SO WHO *DID* STEAL THE ELEPHANT? No one in this town could possibly claim to be an Aharmanite (a Maz-

daysian certainly wouldn't, and an Aharmanite will never truthfully claim to be one), and so no one would claim to worship the same god as an Aharmanite. Everyone in the town would claim to worship the same god as a Mazdaysian. Therefore, if Kushran is Mazdaysian, Shirin answered *yes*, and if Kushran is Aharmanite, Shirin answered *no*. If Shirin answered *yes*, then Omar would know that Kushran is Mazdaysian, hence innocent (since he claimed to be), and Omar couldn't have made a conviction. So, Shirin must have answered *no*, and then Omar knew that Kushran is an Aharmanite, and hence guilty.

88 • AN INTRIGUING MYSTERY This is a good metapuzzle, and I will give the solution in Scheherazade's own words.

"I will first show you, O Auspicious King, that on the basis of the three statements made *before* Omar's question, if *C* is an Aharmanite, then he owns the elephant, and if *C* is a Mazdaysian, then *B* owns the elephant."

"Why is that?" asked the king.

"I will explain: Suppose *C* is an Aharmanite. Then his statement is false, which means that there are not at least two Aharmanites among them; this means *A* and *B* must both be Mazdaysians and their statements are true; and hence, by *A*'s statement, *C* must own the elephant. Now suppose *C* is a Mazdaysian. Then he told the truth, hence *A* and *B* are then Aharmanites, and since *B* then lied, he owns the elephant."

"Very good!" said the king.

"This is as much as can be deduced prior to Omar's question. But then *C* stated who owned the elephant, and although we are not told whom he named, we are told that Omar then knew who the owner was. Now, *C* is either a Mazdaysian or an Aharmanite. If he is Mazdaysian, then *B* owns the elephant, as I have shown, hence *C* being truthful would name *B* as the owner. And so, if *C* is Mazdaysian, he named *B*. But suppose *C* is Aharmanite. Then, as I have shown, *C* owns the elephant; hence *C*, being untruthful, would name either *A* or *B*. So if *C* is an Aharmanite, he named either *A* or *B*. So in either case—whether he is Mazdaysian or Aharmanite—he must have named either *A* or *B*."

"Very good, so far," said the king.

"Now, suppose he named *B*. It could then be that either *C* is Mazdaysian and *B* owns the elephant, or *C* is Aharmanite and the owner of the elephant—and Omar would have no way of knowing which. On the other hand, if *C* named *A*, then he must be an Aharmanite (because if he were a Mazdaysian, he would name *B*, as we have seen) and hence the owner of the elephant. Since Omar *did* know, then *C* must have named *A*, and Omar knew that *C* owned the elephant."

Chapter IX

89 • AT LEAST ONE A statement that works is: I am a Mazdaysian and he is an Aharmanite. If the speaker is telling the truth, then the other one is an Aharmanite, in which case at least one is an Aharmanite. If he is lying, then he is an Aharmanite, and again at least one is an Aharmanite. We note that if the speaker is lying, then the other one could be of either type, and there is no way to tell which. Thus there is no way to tell whether one or both are Aharmanites, and if there is only one, there is no way of telling which one it is. All that can be deduced is that they are not both Mazdaysians.

90 • AT LEAST ONE A statement that works is: Either I am a Mazdaysian or he is an Aharmanite. (We recall that the word *or* is meant to allow the possibility of both.)

If he is a Mazdaysian, then, of course, at least one of them is a Mazdaysian (he is a Mazdaysian and the other one could be either a Mazdaysian or an Aharmanite). Now, suppose he is an Aharmanite. If the other one were also an Aharmanite, then it would be true that *either* the speaker is a Mazdaysian *or* the other one is an Aharmanite. But Aharmanites don't make true statements, and so if he is an Aharmanite, the other one is a Mazdaysian. Thus, in either case, at least one is a Mazdaysian.

91 • ANOTHER TIME A statement that obviously works is: He is an Aharmanite.

92 • HOW MANY OF THEM ARE ROBBERS? If Al-Maamun is lying, then Ubay is an Aharmanite robber, and hence an Aharmanite. If Al-Maamun is telling the truth, then again Ubay is an Aharmanite, since at least one of them is an Aharmanite. This proves that Ubay is definitely an Aharmanite. Therefore he is lying, which means that Al-Maamun is a Mazdaysian who has never committed a robbery. Since Al-Maamun is a Mazdaysian, his statement is true, so Ubay is an Aharmanite who has never committed a robbery. Thus neither one has ever committed a robbery.

93 • HUSSEIN'S ROBBER BAND A statement that would work is: I am an Aharmanite who has never committed a robbery. Obviously, no Mazdaysian would say that, hence the speaker must be an Aharmanite. If he has never committed a robbery, then he really would be an Aharmanite who has never

committed a robbery, but Aharmanites don't make true statements. Therefore, he must have committed a robbery.

94 • THE HOLY MAZDAYSIANS A statement that would work is: I am not a Mazdaysian who has committed adultery. An Aharmanite could not say that (because it is really true that an Aharmanite is *not* a Mazdaysian who has committed adultery), and so the speaker must be a Mazdaysian. Therefore, his statement is true, and so he has never committed adultery.

95 • CASE OF THE STOLEN HORSE If the defendant had answered "Mazda," then Omar would have had no way of knowing whether or not the defendant was guilty, for the defendant could be either an innocent Mazdaysian, a guilty Mazdaysian, an innocent Aharmanite, or a guilty Aharmanite. Therefore, since Omar *did* know, the defendant must have answered "Aharmanite."

Omar reasoned that if the defendant were a Mazdaysian, then it was true that the thief was an Aharmanite, in which case the defendant was innocent. On the other hand, if the defendant was an Aharmanite, then his answer was a lie, which means that the thief was really a Mazdaysian, and hence the defendant was again innocent. Thus, the defendant was innocent, but there is no way to tell whether he was a Mazdaysian or an Aharmanite.

96 • A SOLOMON-LIKE CASE If Safie told the truth, then Zabeide really would have claimed to be the mother, but would have lied (since she is then an Aharmanite), and so in this case, Zabeide is not the mother. On the other hand, if Safie lied, then Zabeide would not have claimed to be the mother, but instead would have said she was not the mother,

and being truthful, she is then not the mother. Thus, regardless of whether Safie told the truth or lied, she is the mother.

97 • A LOGIC PUZZLE I am a lady.

98 • ANOTHER LOGIC PUZZLE I am either a Mazdaysian lady or an Aharmanite man.

99 • ANOTHER LOGIC PUZZLE I am not a Mazdaysian man.

100 • ANOTHER LOGIC PUZZLE I am an Aharmanite man.

101 • WHO IS THE TRAITOR? Ayyib and Isa made incompatible claims, so they can't both be Mazdaysians, and Nowas claimed that all three were of the same type. If Nowas was a Mazdaysian, his claim is true, which would make Ayyib and Isa both Mazdaysians, which we know cannot be. Therefore, Nowas is an Aharmanite, and so, all three are not the same type; therefore either Ayyib or Isa must be a Mazdaysian. This means that either Isa is the traitor (as Ayyib claimed) or Nowas is the traitor (as Isa claimed). In either case, Ayyib cannot be the traitor.

Thus, it was Ayyib who was permitted to leave, and so Isa and Nowas remained. The judge then asked one of these two whether they were of the same type (worshipped the same god) and got an affirmative answer. Could it have been Isa who was asked? No, because we already know that Nowas is an Aharmanite, and no Mazdaysian or Aharmanite would ever claim to be of the same type as an Aharmanite! Therefore, it was Nowas who was asked, and therefore he lied when he answered *yes*, so he and Isa are really of different types, and Isa is a Mazdaysian. Therefore, his claim that Nowas is the traitor was true!

102 • A LOGICAL TANGLE I will give the solution in Scheherazade's own words.

"I will first prove to you, O Mighty King, that *E* and *F* cannot possibly worship the same god, and hence that *G* is lying.

"Suppose *E* is truthful. Then *C* and *D* are both truthful, as *E* claimed. Then *A* and *B* both lie. Then *F* has lied. And so, if *E* is truthful, *F* lies.

"Now, suppose *E* lies. Then *C* and *D* are not both truthful. Hence *A* and *B* do not both lie, and *F* must be truthful. And so if *E* lies, then *F* is truthful. And I have shown that if *E* is truthful, then *F* lies. Therefore, *E* and *F* worship different gods, which proves that *G* worships Aharman.

"Now that we know that *G* lies, we consider *H*'s statement. A truthful man can never make a statement that implies that he worships the same god as the liar *G*. So *H* is certainly a liar, and he does worship the same god as *G*, so the first part of his statement was true. If the second part were also true (that the two defendants are not both guilty) then the whole statement would be true, which cannot be, since *H* is a liar. Therefore, the second clause of his statement was false, and thus both defendants are guilty."

Chapter X

103 • TRICKY NUMBERS Every three-digit number is what Scheherazade humorously called tricky, because writing the number followed by itself is tantamount to multiplying it by 1001. Now, $1001 = 7 \times 11 \times 13$, and so dividing the repeat of the number by 7, 11, and 13 successively is the same as dividing by 1001, which then gives back the original number.

104 • HOW TO DO IT *Method 1:* Start the two hourglasses together. When the 7-minute one runs out, put the egg in the boiling water. Four minutes later, the 11-minute glass runs out. Then turn it over, and when it runs out for the second time, the egg will have boiled for fifteen minutes.

Method 2: Using the above method, one has to wait 22 minutes before eating the egg. One can do the whole job in only 15 minutes by the following more complicated procedure: Drop in the egg immediately, at the same time starting both glasses together. When the 7-minute glass runs out, turn it over. When the 11-minute glass runs out, there will be 4 minutes' worth of sand in the bottom of the 7-minute glass, so turn it over, and when it runs out, 15 minutes will have elapsed.

105 • A VARIANT Again, there are two methods; the first is both simpler and shorter, and is the only one we will give: Drop the egg and start both glasses. When the 4-minute glass runs out, invert it. When the 7-minute one runs out, invert it. A minute later, the 4-minute one runs out; 8 minutes will have elapsed and there will be 6 minutes of sand at the top of the 7-

minute glass and 1 minute of sand at the bottom, so simply invert the 7-minute glass, and a minute later it will have run out and the egg will have boiled for 9 minutes.

106 • CATCH THE BLACK KING Black has the definite advantage, since he can forestall check indefinitely by simply moving and keeping out of a corner square, and on each move, moving to a different colored square than that on which the knight is then resting. On the knight's next move, he will be on the same color as the king, and therefore cannot be checking him.

107 • A GOOD LOGIC PUZZLE A statement that works is: You will not give me either the copper coin or the silver coin. If either the copper coin or silver coin were given, that would falsify the statement, but no coin would be given for a false statement. If no coin at all were given, that would make the statement true, but a true statement gets a coin, so that possibility is out. Therefore, the only possibility is that he gets a gold coin.

108 • AN ARITHMETICAL FACT Many people are quite surprised to find out that the infinite decimal .99999 . . . is *exactly* equal to *1*! This can be seen in several ways: Most people know that the fraction one third is the infinite decimal .33333. . . . Well, multiplying by 3, we get .99999 . . . , and also 3 times one third is *1*.

A second way to see it (which is the way usually presented) is this: Let $x = .9999\ldots$. Multiplying both sides of the equation by 10, we get $10 \times 9.9999\ldots = 9 + .9999\ldots = 9 + x$! Thus $10x = 9 + x$, so $9x = 9$, hence $x = 1$.

Of course, the same method works for any digit other than *9*. For example, let $x = .2222\ldots$. Then $10x = 2.22222\ldots = 2 + .22222\ldots = 2 + x$, hence $9x = 2$, so $x = 2/9$. Thus the value of the infinite decimal $.2222\ldots$ is two ninths (a fact that will be used in the next problem). And similarly, for any digit d from *1* to *9*, the infinite decimal $.ddddd\ldots$ is $d/9$. (For $d = 3$, we get $.33333\ldots = 3/9$, which is the familiar one third.)

To lessen the shock some of you may feel at this point, the following remarks may be helpful: Of course, if you cut off the infinite decimal .99999 . . . at any *finite* place, the result will be less than 1. Thus, the decimal .9999 . . . carried out to a million places is indeed less than one. If, instead, you carry it out to a billion places, it will be closer to 1, but still less than 1. The farther out you carry it, the closer to 1 you will get, but no matter how large a *finite* number of places you carry it out to, the result is always less than 1. Now, what is *meant* by the infinite decimal .99999 . . . ? What is meant is the *limit approached by* the infinite sequence .9, .99, .999, .9999, and so on. And this limit is exactly 1. And that's all that is meant by saying that .9999 . . . equals 1.

109 • THE BOUNCING BALL Let us momentarily forget about the first 180-foot drop and try to calculate the distance traveled after the ball has first hit the ground.

Well, the first up and down is 2/10 of 180; the next is 2/100 of 180; the next is 2/1000 of 180, and so forth. So we must sum the infinite series 2/10 + 2/100 + 2/1000 + . . . , which is simply the infinite decimal .2222 . . . , which we saw in the last problem to be 2/9. And so after first hitting the ground, the ball has traveled 2/9 of 180 feet, which is 40 feet. And so the total distance is 180 + 40, which is 220 feet.

Of course, we are considering a mathematically ideal situation: In the real physical world, the ball wouldn't bounce an infinite number of times before coming to rest! It would probably bounce about 10 or 12 times, and the distance covered would be a fraction of an inch less than 220 feet.

110 • DIVIDING THE SPOILS The better-known method is this: Let's say there are 20 people. Well, one person takes what he considers to be one twentieth of the loot. If all the others agree that he has no more than one twentieth, he walks away with his portion, but if anyone believes that he has more than one twentieth, the portion is passed on to him with instructions to put back into the pot enough so that he, the second person, feels that he is holding one twentieth. If no one else objects, he walks off with this shaved-down portion; otherwise, the portion is passed to a third person who shaves it down to what *he* regards as one twentieth, and so forth. Sooner or later, one person must have what he regards as one twentieth while no one else feels that he has more than one twentieth, so he is content to walk off with his portion and no one objects to his doing so. Then the problem reduces to 19 people dividing the remaining spoils, and they go through the same process as before (only with one nineteenth instead of one twentieth) thus reducing the problem to 18 people dividing the spoils, and so on down the line.

The second method (which seems to me less well known) is this: First, two people divide the entire pot between them into what each considers his fair share; this is done by the familiar device of one person dividing the pot into what he considers two equal parts and having the other one choose. Then each of the two divides his portion into what he considers three equal parts, and then a third person of the group picks his favorite third of the first person's portion and his favorite third of the second person's portion. Now, each of the three then feels that he has at least one third of the pot. Then each of the three divides his portion into what he considers four equal parts, and a fourth person then picks one part from each of the three, and so we now have four people, each convinced that he has at least one fourth of the pot. And this process is continued—until finally the twentieth person picks what he considers at least one twentieth from each of the other 19 people.

111 • A PARADOX The fallacy is this: Consider the statement: If Box B is empty, then the chances are 50–50 that the coin is in Box A. Although this has the grammatical form of an *if / then* statement, it is not really one; it is really a statement of *conditional probability* and simply means that in all those possibilities in which Box B is empty, in half of them the coin will be in Box A. True enough! Also, in half the possible cases in which Box C is empty, the coin will be in Box A. But there is no inconsistency in these two statements.

CHAPTER XI

112 • SCRAMBLED LABELS It is impossible that only one is wrong, for if two are right, the third one must also be right.

113 • MORE SCRAMBLED LABELS One can do this by opening only one drawer! You open the drawer in the chest labeled *"E R."* Suppose you find an emerald. Since you know that the label is wrong, then your chest must really contain both emeralds. Then, the chest labeled *"R R"* cannot really

contain 2 rubies (since the label is wrong), nor 2 emeralds (since the chest you have picked contains 2 emeralds), and so the chest labeled *"R R"* must be the mixed chest, and hence the chest labeled *"E E"* contains both rubies.

Of course, if you originally picked a ruby from the chest labeled "*E R*," the argument is symmetric—that is, the chest labeled *"E E"* is the mixed chest and the one labeled *"R R"* really contains 2 emeralds.

114 • A SCRAMBLED LABEL METAPUZZLE I will show that if the third man knew what he had, then there are two possible solutions; whereas if he didn't know, then there is only one. And so, the only way you could obtain the solution by being told what the third man said is by being told that the third man said that he didn't know.

Now for the details: Since the first man found two emeralds and knew what he had, then he either saw the label *"E E E"* and knew that he really had *E E R*, or he saw *"E E R"* and knew that he really had *E E E*.

Since the second man found an emerald and a ruby and knew what he had, then he either saw the label "*E E R*" and had *E R R*, or he saw "*E R R*" and had *E E R*.

The third man found two rubies, and there are now two possibilities:

> *Case 1:* He knew what he had, hence he either saw "*E R R*" and had *R R R*, or saw "*R R R*" and had *E R R*.
>
> *Case 2:* He didn't know, and in this case, he saw either "*E E E*" or "*E E R*."

In Case 1, there are two possible solutions. The first is:

First man saw "*E E E*" and had *E E R*.
Second man saw "*E E R*" and had *E R R*.
Third man saw "*E R R*" and had *R R R*.
Fourth man saw "*R R R*" and had *E E E*.

The second possible solution is:

First man saw "*E E R*" and had *E E E*.
Second man saw "*E R R*" and had *E E R*.
Third man saw "*R R R*" and had *E R R*.
Fourth man saw "*E E E*" and had *R R R*.

Now, suppose Case 2 holds. We will see that there is then only one possible solution: Suppose the first man saw "*E E E*" (which we will see is impossible). Then he really has *E E R*, hence the second man couldn't have *E E R*, so he must have *seen* "*E E R*" and really had *E R R*. Then, the third man couldn't have seen "*E E E*" (since the first man did), nor "*E R R*" (since the second man saw "*E R R*"), which violates Case 2. And so, in Case 2, the first man *didn't* see "*E E E*." Therefore, he saw "*E E R*" and really had *E E E*. The second man didn't see "*E E R*," so he saw "*E R R*" but really had *E E R*. And the third man didn't see "*E E R*" (the first man did), so he saw "*E E E*." This leaves the remaining label "*R R R*" on the fourth chest, and since it was wrong, *R R R* was really the content of the third chest.

Thus, the solution is that the four chests, in order, contained *E E E*, *E E R*, *R R R*, *E R R*, and the labels were, in order, "*E E R*," "*E R R*," "*E E E*," "*R R R*."

115 • WHAT ARE THEIR AGES? The following are the only triples whose product is 36 (their sum is written alongside).

1, 1, 36—38	1, 6, 6—13
1, 2, 18—21	2, 2, 9—13
1, 3, 12—16	2, 3, 6—11
1, 4, 9—14	3, 4, 3—10

The only two triplets having identical sums are 1, 6, 6 and 2, 2, 9, the sum of which is 13. Now, if Iskandar was any age other than 13, he would have known the correct triple when told the sum. (We are of course assuming that he knew his own age.) But he didn't know at this point, and therefore his age must be 13, and he knew at that point that the triple was either 1, 6, 6 or 2, 2, 9, but he had no way of telling which. But then when he was told that the oldest was at least one year older than the other two, that ruled out 1, 6, 6, and so he then knew that the son was 9 and his sisters were each 2.

116 • WHAT TYPE IS BULIKAYA? I will give the solution in Scheherazade's own words.

"True, Your Majesty, I did not tell you what either Ayn Zar or Bulikaya said, but I will first show you that if Ayn Zar had answered *no*, then Omar would have known what Bulikaya was without having to ask a second question. Suppose Ayn Zar had answered *no*, then Ayn Zar would in effect be claiming that he and Bulikaya were both Aharmanites, which is possible, but only if he is an Aharmanite and Bulikaya is a Mazdaysian. So, if Ayn Zar had answered *no*, Omar would have concluded that Bulikaya was a Mazdaysian and wouldn't have had to ask a second question. But Omar did ask a second question, hence Ayn Zar must have answered *yes*. At this point, all Omar knew was that it was *not* the case that Ayn Zar was an Aharmanite and Bulikaya a Mazdaysian (for otherwise, Ayn Zar would have answered *no*). And so Omar then knew that one of the following three possibilities held:

Case 1: Ayn Zar is a Mazdaysian and Bulikaya an Aharmanite.

Case 2: Ayn Zar is a Mazdaysian and Bulikaya a Mazdaysian.

Case 3: Both are Aharmanites.

"Next, Omar asked Bulikaya whether Ayn Zar had told the truth—in other words, whether Ayn Zar was a Mazdaysian. If Case 1 holds, then Bulikaya would answer *no*. If either Case 2 or Case 3 holds, Bulikaya would answer *yes*. So if Bulikaya had answered *yes*, then Omar couldn't have known whether Case 2 or Case 3 held, hence he wouldn't have known what Bulikaya was. But Omar *did* know, hence Bulikaya must have answered *no*, and Omar then knew that Case 1 was true, and hence Bulikaya was an Aharmanite.

Chapter XII

117 • THE FIRST TEST The key to all these tests is to find a number *k* such that for any number *x* the number *k1x2* will bring back *x1x2*. Then, taking for *x* the number *k* itself, we see that *k1k2* brings back *k1k2*—the very same number! Thus finding such a number enables the suitor to pass the first test. Such a number *k* has other uses as well, as we will see. We will

call such a number *k* a *key* number. Now the problem is to find a key number.

One key number is 475364. To see why, it will be convenient to denote the *reverse* of a number *x* by the symbol $\overleftarrow{x}$. Now, consider any number *x*. The following facts can be successively seen to hold:

1. *1x2* gets back *x*.
2. Therefore, *41x2* gets back $\overleftarrow{x}$.
3. Therefore, *641x2* gets back $1\overleftarrow{x}$.
4. Therefore, *3641x2* gets back $1\overleftarrow{x}1\overleftarrow{x}$.
5. Therefore, *53641x2* gets back $\overleftarrow{x}1\overleftarrow{x}$.
6. Therefore, *753641x2* gets back $2\overleftarrow{x}1\overleftarrow{x}$.
7. Therefore, *4753641x2* gets back *x1x2* (the reverse of $2\overleftarrow{x}1\overleftarrow{x}$).

Thus, 475364 is a key number, and so a number that brings back itself is 47536414753642, as the reader can directly verify. This ensures passing the first test.

There is in fact an infinite quantity of key numbers, for if *k* is a key number, so is *44k*, and so is *4444k*, and indeed any even number of 4s followed by *k*. (In general, if number *x* brings back *y*, so will *44x*, since the reverse of the reverse of *y* is *y* itself.)

118 • THE SECOND TEST For any key number *k*, the number *3k13k2* will bring back its own repeat, because *k13k2* brings back *3k13k2*, so *3k13k2* brings back the repeat of *3k13k2*.

In particular, taking 475364 for *k*, the number 3475364134753642 brings back its repeat.

119 • THE THIRD TEST For any key number *k*, the number *4k14k2* brings back its own reverse (since *k14k2* brings back *4k14k2*). Taking 475364 for *k*, we get the solution 4475364144753642, which has 16 digits. However, we can cancel both the double 4s, getting 753641753642 as a shorter solution (12 digits), which the reader can verify directly, or see that it must be so by the following argument:

In constructing a key number, we saw that the number 75364 has the property that for any number *x*, *753641x2* gets back $2\overleftarrow{x}1\overleftarrow{x}$. Well, for *any* number *n* with this property, *n1n2* will get back $2\overleftarrow{n}1\overleftarrow{n}$—the reverse of *n1n2*.

120 • THE FOURTH TEST It easily follows from the rules that 536 has the property that for any number *x*, the number *5361x2* brings back *x1x*. In particular, taking 536 for *x*, 53615362 will bring back 5361536.

121 • THE FIFTH TEST What is needed is a number *y* that brings back *1y2*. Then *1y2* also brings back *y*, so the suitor could send either *y* or *1y2*, and thus get back the other, which when sent back to the princess, will bring him back the first.

There are several ways to find such a number *y*. Here is one: We saw in Step 6 of the constructions of a key number that for any number *x*, the number *753641x2* brings back $2\overleftarrow{x}1\overleftarrow{x}$; therefore, *7753641x2* brings back $22\overleftarrow{x}1\overleftarrow{x}$, and *47753641x2* brings back *x1x22*, and so *647753641x2* will bring back *1x1x22*. To reduce clutter, let *b* be the number 64775364. Thus *b1x2* brings back *1x1x22*, for any *x*, and so, in particular, *b1b2* brings back *1b1b22*, and so we take *y* to be *b1b2*, and so *y* brings back *1y2*. Thus, we can take *y* to be 647753641647753642. And so the suitor can send either this *y* or *1y2*.

122 • THE SIXTH TEST We want a number *n* that will bring back *41n2*, which in turn brings back the reverse of *n*.

Let *c* be the number 4775364. It has the property that for any number *x*, the number *c1x2* brings back *x1x22*, as is easily verified. Incidentally, the number *b* of the last problem is *6c*. Then, taking *41c* for *x*, we see that *c141c2* brings back *41c141c22*, and so we take *n* to be *c141c2*, and *n* brings back *41n2*. Since *c* is the number 4775364, then our solution is 477536414147753642.

123 • THE SEVENTH TEST The trick is to find a number *y* that gets back *2y1*. If the suitor then sends *1y2*, the princess will send back *y*, which he will return to her, and she will send back *2y1*, and he will pass the test.

To find such a number *y*, we need a number *d* with the property that for all *x*, the number *d1x2* gets back *2x1x21*. Then *d1d2* will get back *2d1d21*, and so we take *y* = *d1d2*. Now, such a number *d* is 46747536 (as is easily verified). Thus we take *y* = 467475361467475362, and so the suitor should send *1y2*, which is 14674753614674753622.

124 • THE EIGHTH TEST There are several ways of going about this: One way is to find a number *x* that brings back *22x12x11*. He then reverses *22x12x11* and sends $11\overleftarrow{x}21\overleftarrow{x}22$ to the princess. She returns him $1\overleftarrow{x}21\overleftarrow{x}22$. He then "halves" it, sending her $1\overleftarrow{x}2$. She then sends back $\overleftarrow{x}$, and he passes!

How to find such an *x*? Well, we first need a number *n* with the property that for every number *y*, the number *n1y2* brings back *22y1y212y1y211*. Then *n1n2* will bring back *22n1n212n1n211*, and we then take *x* = *n1n2*. How do we obtain such a number *n*? Well, we can get from *y* to 22*y*1*y*212*y*1*y*211 via the following steps:

y
$\overleftarrow{y}$ (reverse, using 4)
$1\overleftarrow{y}$ (prefix 1, using 6)
$1\overleftarrow{y}1\overleftarrow{y}$ (repeat, using 3)
$\overleftarrow{y}1\overleftarrow{y}$ (erase, using 5)
$2\overleftarrow{y}1\overleftarrow{y}$ (prefix 2, using 7)
$12\overleftarrow{y}1\overleftarrow{y}$ (prefix 1, using 6)
$212\overleftarrow{y}1\overleftarrow{y}$ (prefix 2, using 7)
$y1y212$ (reverse, using 4)
$y1y212y1y212$ (repeat, using 3)
$212\overleftarrow{y}1\overleftarrow{y}212\overleftarrow{y}1\overleftarrow{y}$ (reverse, using 4)
$12\overleftarrow{y}1\overleftarrow{y}212\overleftarrow{y}1\overleftarrow{y}$ (erase, using 5)
$112\overleftarrow{y}1\overleftarrow{y}212\overleftarrow{y}1\overleftarrow{y}$ (prefix 1, using 6)
$y1y212y1y211$ (reverse, using 4)
$2y1y212y1y211$ (prefix 2, using 7)
$22y1y212y1y211$ (prefix 2, using 7)

Thus we take *n* to be 774654347675364, and so we take *x* to be 77465434767536417746543476753642.

CHAPTER XIII

125 • THE FIRST VERSION According to this version, Scheherazade asked: Will you either answer *no* to this question or spare my life?

What Scheherazade is asking here is whether at least one of the following alternatives holds:

1. The king will answer *no*.
2. The king will spare her life.

To answer *yes* to this question is to affirm that at least one of the two alternatives (1 or 2) holds. To answer *no* is to claim that neither alternative holds. But if the king answers *no*, then Alternative 1 *does* in fact hold, and so *no* is therefore a false answer! In order to answer truthfully, the king must answer *yes*. He thus must affirm that at least one of the two alternatives holds. But when he answers *yes*, Alternative 1 does not hold, so it must be that if the king's answer is truthful, Alternative 2 does hold! Or, we can put it this way: If he answers *yes* and doesn't spare her life, then he is affirming that at least one of the two alternatives holds, whereas, in fact, neither one then holds, and hence his answer wouldn't be truthful.

Thus, the only possibility is that the king answers *yes* and spares her life.

An alternative way of putting the question is: If you answer *yes*, will you then spare my life?

126 • THE SECOND VERSION A question that works is: Will you answer *no* and take my life?

Obviously, *yes* cannot be a correct answer to that question (regardless of whether the king takes her life or spares it), so the king must answer *no*. By answering *no*, the king is denying that both of the following propositions are true:

1. The king answers *no*.
2. The king takes her life.

In other words, by answering *no*, the king is claiming that at least one of the propositions (1 or 2) is false, but it can't be the first since the king answered *no*, so it must be the second.

Thus, the king is, in effect, asserting that Proposition 2 is false, which means he must spare her life! Stated otherwise, if the king answers *no* and takes her life, then both propositions are true, which he shouldn't have denied by saying *no*.

127 • THE THIRD VERSION A question that works is: Will you either answer *yes* and spare my life, or answer *no* and take my life? Thus, Scheherazade is asking whether one of the following alternatives holds:

1. The king answers *yes* and spares her life.
2. The king answers *no* and takes her life.

Suppose he answers *yes*. Then he is claiming that one of those alternatives holds, which obviously can't be the second, so it must be the first. Thus, the king must spare her life if he answers *yes*.

Now, suppose he answers *no*. Then he is denying both alternatives. The only way he can truthfully deny the second is by sparing her life (because if he takes her life then it is true that he answered *no* and took her life, so it shouldn't be denied)!

Thus the king has the option of answering *yes* or answering *no*, but in either case, he must spare her life.

An equivalent form of the question is: Will you answer *yes* if and only if you spare my life?